# 7th International Symposium on Big Data and Applied Statistics (ISBDAS 2024)



Journal of Physics: Conference Series Volume 2813

Beijing, China
8-10 March 2024

ISBN: 979-8-3313-0287-0
ISSN: 1742-6588

**Printed from e-media with permission by:**

Curran Associates, Inc.
57 Morehouse Lane
Red Hook, NY 12571

**Some format issues inherent in the e-media version may also appear in this print version.**



Institute of Physics
Dirac House, Temple Back
Bristol BS1 6BE UK

Phone: 44 1 17 929 7481
Fax: 44 1 17 920 0979

techtracking@iop.org

**Additional copies of this publication are available from:**

Curran Associates, Inc.
57 Morehouse Lane
Red Hook, NY 12571 USA
Phone: 845-758-0400
Fax: 845-758-2633
Email: curran@proceedings.com
Web: www.proceedings.com

# TABLE OF CONTENTS

Journal of Physics: Conference Series **2813** (2024) 011001 doi:10.1088/1742-6596/2813/1/011001

# Preface

With the rapid development of information technology, big data has become an important force to promote social progress and scientific and technological innovation. At the same time, applied statistics, as a bridge connecting data and knowledge, has become increasingly prominent in its central role in big data analysis. Against this background, we have welcomed the 2024 7$^{th}$ International Symposium on Big Data and Applied Statistics (ISBDAS 2024) held virtually from March 8$^{th}$ to 10$^{th}$, 2024 in Beijing, China.

This Symposium aimed to bring together experts and scholars, engineers and technologists, and technology developers in the field of big data and applied statistics from all over the world to share scientific research results, discuss cutting-edge technologies, and promote academic progress and industrial cooperation.

The topics of this Symposium are broad and in-depth, covering all aspects of big data and applied statistics. We have invited renowned scholars and industry experts from all over the world to conduct in-depth exchanges and discussions on the latest research results, technology development trends, and industrial application prospects of big data and applied statistics. During the Symposium, participants listened to exciting presentations from different fields, had in-depth exchanges with their peers, and jointly promoted the development of big data and applied statistics and their relevant areas.

The compilation of the proceedings of ISBDAS 2024 is a comprehensive presentation of the academic results of this Symposium. We received many high-quality submissions from all over the world, and after a rigorous review process, we finally included these papers with innovative and practical value. These papers not only cover multiple research directions in big data and applied statistics, but also demonstrate the latest research results and technical applications. We believe that the publication of these papers will have a positive impact on the research and practice of big data and applied statistics.

In the process of compiling the proceedings, we have received great support and help from many experts, scholars and reviewers. They carefully reviewed each paper and made valuable comments and suggestions to ensure the quality and level of the contributions. We would like to express our heartfelt thanks to all the experts and scholars who participated in reviewing and submitting the papers!

Finally, let us once again thank all the experts and scholars, engineers and technicians, and technology developers who participated in this Symposium and contributed to the proceedings. Thank you for your hard work and contributions, and let's look forward to the bright future of big data and applied statistics together!

The Committee of ISBDAS 2024

Journal of Physics: Conference Series **2813** (2024) 011001 doi:10.1088/1742-6596/2813/1/011001

# Committee Member

**Honorary Chair**
Ljiljana Trajkovic, Simon Fraser University, Canada

**Conference Chair**
Philippe Fournier-Viger, Shenzhen University, China
Liang Yao, Indiana University-Purdue University Indianapolis, USA
Gyu Myoung Lee, Liverpool John Moore, UK

**Technical Program Committee Chair**
Yudong Zhang, University of Leicester, UK
Xiangjie Kong, Zhejiang University of Technology (ZJUT), China
Pascal Lorenz, University of Haute-Alsace, France
Tianrui Li, Southwest Jiaotong University, China
Yan Yang, Southwest Jiaotong University, China
Harris Wu, Old Dominon University, USA
Wong Kuan Yew, Universiti Teknologi Malaysia (UTM), Malaysia

**Publication Chair**
Haw Su Cheng, Multimedia University, Malaysia
Parikshit N. Mahalle, Vishwakarma Institute of Information Technology, India

**Technical Program Committee Member**
Yueting Zhuang, Zhejiang University, China
Demetri Terzopoulos, University of California at Los Angeles, USA
Guiguang Ding, Tsinghua University, China
Nuno Vasconcelos, University of California at San Diego, USA
Xiaojin (Jerry) Zhu, University of Wisconsin-Madison, USA
Alan Jeffrey Giacomin, Queen's University, Canada
Zuohuan Zheng, Chinese Academy of Sciences, China
Zhen Wang, Dalian University of Technology, China
Jianguo Liu, Imperial College London, UK
Ming Tai-seale, University of California at San Diego, USA
Paul Wen, University of Southern Queensland, Australia
Zygmunt J. Haas, Cornell University, USA
Malik Zawwar Hussain, University of the Punjab, Lahore, Pakistan
Jingshan Huang, The University of South Alabama (South), USA
Jawahar Thakur, Himachal Pradesh University, India
Qingzhong Liu, The Sam Houston State University, USA
Mariofanna Milanova, University of Arkansas Little Rock, USA
Qiang Yang, Zhejiang University, China
Alireza Vali Pour Baboli, Technical and Vocational University, Iran
Suraiya Jabin, Jamia Millia Islamia, India
Zoran Gajic, University of New Jersey, USA
Ephraim Suhir, Portland State University, USA
Afaq Ahmad, Sultan Qaboos University, Oman
Kannimuthu Subramanian, Karpagam College of Engineering, India
Rao Li, The University of South Carolina Aiken, USA
Tieyong Zeng, The Chinese University of Hong Kong (CUHK), Hong Kong, China
M-Tahar Kechadi, University College Dublin, Ireland
Yilun Shang, Northumbria University, UK
Pavel Loskot, Zhejiang University, China
Naragain Phumchusri, Chulalongkorn University, Thailand
GAJENDRA Sharma, Kathmandu University, Nepal
Thomas Lee, University of Illinois at Chicago, USA
Fernanda Otilia Figueiredo, Universidade do Porto, Portugal

Journal of Physics: Conference Series **2813** (2024) 011001 doi:10.1088/1742-6596/2813/1/011001

Tien-Ying Kuo, National Taipei University of Technology, Taiwan, China
Khondker Shajadul Hasan, University of Houston-Clear Lake, USA
Cheng Chin, Newcastle University, Singapore
Malinka Spasova Ivanova, The Technical University of Sofia, Bulgaria
Zhiyu Jiang, Northwestern Polytechnical University, China
Dariusz Jacek Jakobczak, Koszalin University of Technology, Poland

Journal of Physics: Conference Series **2813** (2024) 011002 doi:10.1088/1742-6596/2813/1/011002

# Peer Review Statement

All papers published in this volume have been reviewed through processes administered by the Editors. Reviews were conducted by expert referees to the professional and scientific standards expected of a proceedings journal published by IOP Publishing.

- **Type of peer review:** Single Anonymous
- **Conference submission management system:** Morressier
- **Number of submissions received:** 80
- **Number of submissions sent for review:** 64
- **Number of submissions accepted:** 27
- **Acceptance Rate (Submissions Accepted / Submissions Received × 100):** 33.8
- **Average number of reviews per paper:** 3
- **Total number of reviewers involved:** 7
- **Contact person for queries:**
  **Name:** Xuexia Ye
  **Email:** xx.ye@keoaeic.org
  **Affiliation:** AEIC Academic Exchange Information Centre

Journal of Physics: Conference Series **2813** (2024) 012001 doi:10.1088/1742-6596/2813/1/012001

# Adaptive spatial attention dual-branch fishing boat detection network

**Jiaxuan Yang* and Xiang Liu**

Dalian Maritime University, College of Navigation, Dalian, Liaoning,116026, China

*Corresponding author's e-mail: yangjiaxuan@dlmu.edu.cn

**Abstract:** Aiming at the harbor environment, the target detection accuracy of fishing vessels is low, and it is prone to the problems of fishing vessel misdetection and omission detection. In this paper, we propose a fishing vessel target detection algorithm called ASDNet based on YOLOX. Firstly, an Adaptive Spatial Attention Module (ASAM) was designed and used to improve the detection of fishing vessel targets; secondly, a two-branch backbone network was designed for multidimensional fishing vessel feature extraction. Meanwhile, a bilateral enhanced fusion strategy (BFFS) is designed to fuse the branch features to improve the characterization ability of the network; finally, the loss function is improved by introducing the Focal-CIOU loss bounding box loss function to reduce the effects of the detection position deviation of the fishing vessel target and the overlap of the vessel hull to improve the detection performance. The above methods are validated using the homemade fishing vessel dataset, and the results show that the precision rate (P) and recall rate (R) are greatly improved. The average precision rate (mAP@50-95) value reaches 80.25%, which is 2.39% higher than that of the 77.86% of the YOLOX. It significantly improves the precision of the detection, meets the requirements of the performance of the target detection of the fishing vessel, and has certain practical significance in engineering.

## 1. Introduction

To ensure the safe navigation of ships, some port areas have set up no-navigation zones for fishing vessels. However, as the sea production of fishing vessels is open, the shore-based management personnel are not strong enough and the means are relatively backward, so they are unable to control the dynamics of fishing vessels in real-time, not to mention that they cannot timely detect and eliminate such irregularities as private fishing in the no-navigation zones during the fishing season, and there is a blind spot in the supervision of these potential safety hazards. In the early days, the task of monitoring fishing vessels in the port area was mainly realized through AIS equipment, but some fishing vessels chose to close the AIS equipment to avoid the supervision of the maritime regulatory authorities, and even the phenomenon of modifying the AIS information exists. Faced with this situation, a single reliance on the AIS maritime supervision system will not be able to complete the effective monitoring of fishing vessels, which also poses a threat to the safe navigation of ships in the harbor area. Therefore, port managers have been looking for a solution to monitor fishing vessels without relying on AIS to ensure the safe navigation of vessels in the port area.

To solve the over-reliance on the AIS system for fishing vessel monitoring in the harbor area, this paper proposes a method combining deep learning with visual sensors to assist in supplementing the monitoring of fishing vessels in port areas. Searching the existing literature, there are fewer studies on fishing vessel monitoring based on visual sensors, and most of them are for cargo vessel monitoring. Up to now, research on cargo ship target detection algorithms based on RGB images can be roughly divided into two categories: the first category combines the target detection algorithm with the attention mechanism, and this type of approach starts from the perspective of solving the monitoring of small-scale cargo ships, e.g., YOLO-Ship[1], YOLOv7-sea[2] with improved YOLOv4[3]. For the small-scale cargo ship target, in this paper, the ratio of the pixel area of ship target instances to the pixel area of the image in RGB images ranges from 0.08% to 0.58%. Regarding this issue, most existing studies employ a uniform attention mechanism in the backbone network. Since most of the cargo ships navigate within the customized channel in the port area when the visual sensors are used to capture the cargo ships, they are presented as uniform scale ship targets, and this scheme has a significant improvement in the monitoring effect of the cargo ships. However, due to the flexibility and mobility of fishing vessels, their navigation status changes with the fishing status, and their visual distance from the vision sensor is random and uncontrollable. This will result in the RGB image captured by the visual sensor, with a change in the visual distance between the fishing vessel and the visual sensor. The fishing vessel is presented as a multi-scale target in the RGB image, and at this time, there will be some leakage and misdetection. In this paper, a multi-scale fishing vessel target is defined as the area of the target instance to the area of the image in the RGB image between 0.08% and 28.0% for the same fishing vessel and the visual sensor at different view distances. It is difficult to perceive the multi-scale fishing vessel target better using the unified attention mechanism, and it will be interfered with by non-correlated information. The second class of methods uses better-performance backbone network-optimized target detection algorithms, such as ST-YOLOA[4], and GGT-YOLO[5] with PAG-YOLO[6]. The role of the backbone network in the target detection algorithm is to extract and compress the features such as shape contour, position, and hull color of the target instance in the RGB image. In the monitoring of cargo ships, due to its larger hull, the shape is easy to distinguish, in the sea and sky background has a more obvious contour and ship shape features, and the backbone network can be more complete compression and extraction of its features. Fishing vessels are generally small in tonnage and have complex and diverse ship types, strong concealment, and other characteristics. Their outline shape is vague, the hull is easy to integrate with the sea and sky background, and the outline shape of the fishing vessel target and ship features in the compression of the outline shape is very easy to lose, which results in the monitoring of the location of the vessel cannot be captured information. Even with the more advanced backbone network, it is still impossible to prevent the loss of positional features, and a large number of omissions and misdetections will occur at this time.

To compensate for the problems of visual sensors in fishing vessel monitoring, this paper proposes a two-branch target detection network (ASDNet) algorithm with an adaptive spatial attention mechanism, which uses YOLOX as the Baseline and improves the performance of monitoring multi-scale fishing vessel targets on the sea surface. Specifically, to enhance the ability to perceive multi-scale fishing vessel targets in RGB images, in this paper we design an adaptive spatial attention module; to solve the problem of the loss of positional contour features of the fishing vessel targets, a spatial positional network branch is introduced into YOLOX to obtain the rich information of positional contour features in the RGB images. Meanwhile, a bilateral feature enhancement module is used to combine semantic network branching features and spatial location network branching features by using the bilateral feature enhancement module.

## 2. Methodology

### *2.1 Network structure*

Here, the ASDNet network is optimized for detecting and recognizing fishing vessel targets, using YOLOX as the benchmark algorithm (as shown in Figure 1). Unlike the YOLOX network, which uses CSPDarkNet as the backbone network, firstly, we replace CSPDarkNet with Inverted-CSPDarkNet. Meanwhile, the adaptive spatial attention is integrated into the Inverted-CSPDarkNet network to improve the network's ability to perceive multi-scale fishing vessel targets; and then, a light-weighted spatial position network branch focuses on the rich position information in stage 2; finally, a bilateral feature enhancement strategy is utilized to fuse the two-branch features.

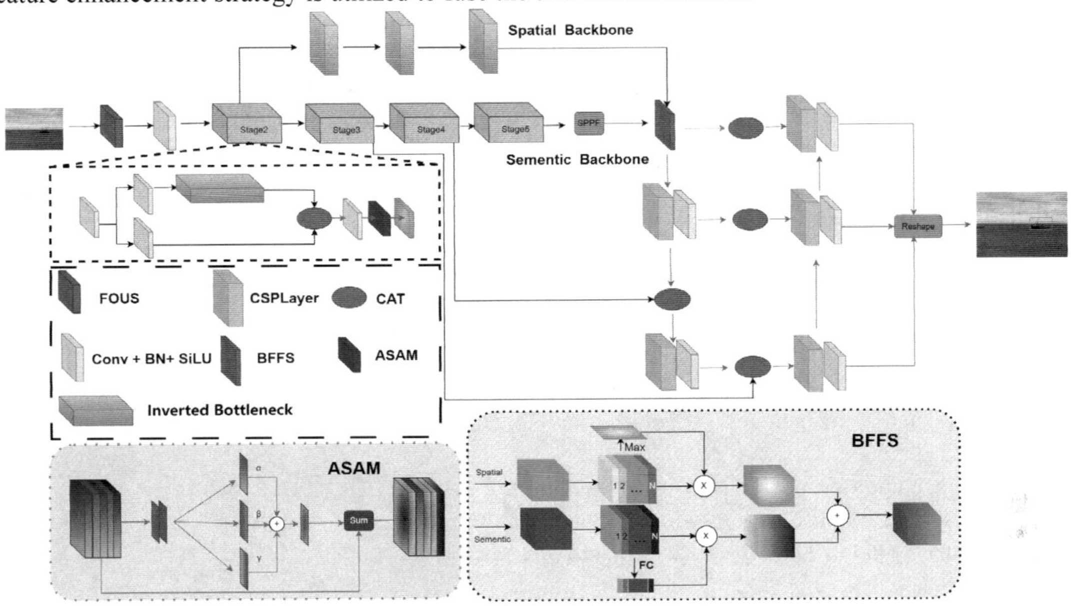


**Figure 1.** The structure of ASDNet.

*2.2 Dual-backbone network*

Here, we propose a two-branch backbone architecture consisting of a spatial location network branch and a semantic network branch (as shown in Figure 2) to compensate for the loss of small-scale fishing vessel target location features caused by a single semantic network branch. In this paper, the spatial location network branch is constructed by utilizing the rich location information in stage 2, which focuses on the target location information through the location feature extractor; for the semantic network branch, we still adopt the CSPDarknet structure to extract image semantic features, and at the same time, we draw on the idea of Inverted residual[7] to construct the Inverted-CSPDarknet branch. Inverted-CSPDarknet is used for semantic network branching.

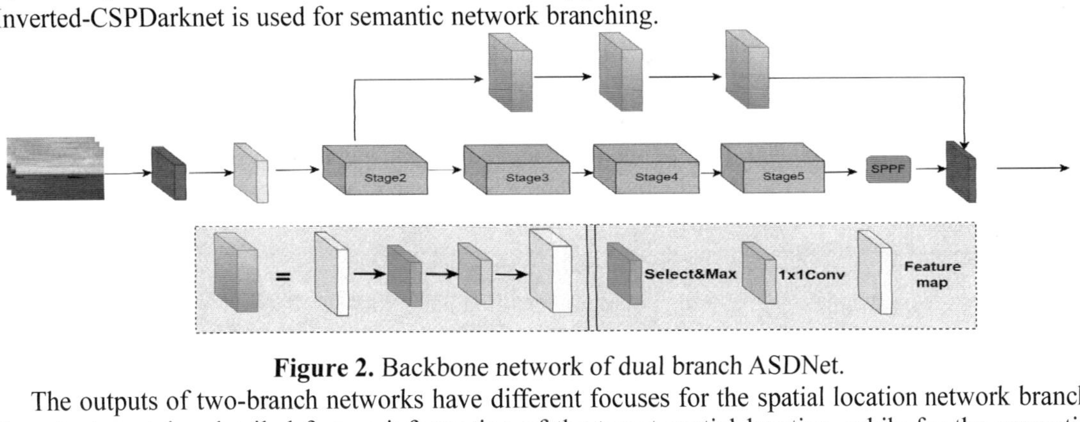


**Figure 2.** Backbone network of dual branch ASDNet.

The outputs of two-branch networks have different focuses for the spatial location network branch its output contains detailed feature information of the target spatial location, while for the semantic network branch, its output contains rich semantic feature information[8]. An important task is to fuse the two, and this paper proposes an efficient bilateral feature enhancement fusion strategy (BFFS), as shown in Figure 1, which can realize the effective combination of spatial detail information and rich semantic information.

*2.3 Adaptive spatial attention mechanism*

Here, adaptive spatial attention (ASAM) is designed for the multi-scale presentation of fishing boat targets in RGB images, which is designed to adaptively model the input feature maps from the multi-scale receptive fields for solving the perceptual ability of fishing boat targets of different scales under different viewing distances[9]. The adaptive spatial attention module in this paper uses filters with different receptive fields for the same attentional weights to model their multiscale attention and introduces learnable weights that enable the network to learn by itself to get the most favorable attentional pattern for the scene.

*2.4 Loss function*

The loss functions in YOLOX include categorization, position, and confidence. For the position loss in ASDNet, we use the CIOU loss function[10] to determine the position of the detection frame of the target; for the category loss, we use Focal loss[11] for confidence fitting. Where the Focal loss function is calculated as expressed in Eq. (1); for CIOU loss is calculated as shown in Eqs. (2) ~ (5).

$$FL(p_t) = -\alpha_t (1-p_t)^{\gamma} \log(p_t) \tag{1}$$

where $p_t$ is the prediction probability, αt represents the balance coefficient of each category, by which the network's focus on different categories of targets can be modulated, and (1 - pt) γ is the modulation factor, which adjusts the network's focus on the difficult-to-split samples.

$$IOU = \frac{A \cap B}{A \cup B} \tag{2}$$

$$\alpha = \frac{v}{(1-IOU)+v} \tag{3}$$

$$v = \frac{4}{\Omega^2}(\arctan\frac{w^{gt}}{h^{gt}} - \arctan\frac{w}{h})^2 \tag{4}$$

$$CIOU_{loss} = 1 - IOU + \frac{d^2}{c^2} + \partial v \tag{5}$$

where A represents the predicted bounding box during the network training process, and B represents the annotated bounding box in the image. α is the weighting parameter for positively predicted boxes used to measure aspect ratio, where $w^{gt}$ and $h^{gt}$ denote the width and height of the annotated box, and w and h represent the width and height of the predicted box, respectively. d is the straight-line distance between the center points of the predicted and annotated boxes, while C is the diagonal distance between the minimum bounding rectangles of the predicted and annotated boxes.

## 3. Experimental and analysis

*3.1 Dataset construction*

Here, we use Seaships datasets, which contain 7000 images, and a fishing vessel dataset was produced using the fishing vessel part of this dataset. The examples of fishing boats of different scales in the fishing boat dataset are statistically examined and the results are shown in Figure 4(a). We can see that the number of small and medium-scale fishing boats is small. To make the dataset as balanced as possible for all scales of fishing boats, this paper will be four large and medium-scale fishing boat images spliced and then scaled down to the size of a single image used to simulate the target of the medium and small-scale fishing boats (as shown in Figure 3), and finally get the Fishing-ours dataset, which is shown in Figure 4 (b) for all categories.

**Figure 3.** We use large-scale and middle-scale fishing vessel simulations to simulate middle-scale and small-scale fishing vessel examples, respectively, with the simulated middle-scale on the left and the simulated small-scale on the right.

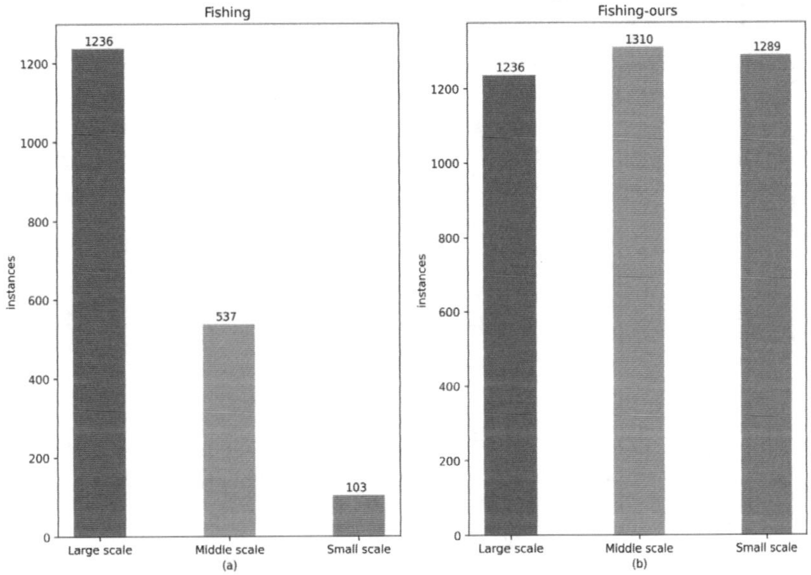


**Figure 4.** Analysis of the fishing dataset.

*3.2 Evaluation metric*

In this paper, Precision (P), Recall (R), and mean Average Precision (mAP) are used to evaluate the model's performance. Their equations are as follows:

$$\mathrm{m}AP = \frac{1}{n}\sum_{i=1}^{n} AP_i \tag{6}$$

$$AP = \int_0^1 P(k)dR(k) \tag{7}$$

$$P = \frac{TP}{TP + FP} \tag{8}$$

$$R = \frac{TP}{TP + FN} \tag{9}$$

where TP represents the number of objects correctly identified as vessels; FP represents the number of background objects incorrectly identified as vessel targets; FN represents the number of vessel targets not detected; C denotes the number of vessel categories; n represents the number of IOU thresholds; k represents the IOU threshold; P(k) denotes the precision at the threshold k, R(k) denotes the recall at the threshold k.

Journal of Physics: Conference Series **2813** (2024) 012001 doi:10.1088/1742-6596/2813/1/012001

### *3.3 Ablation experiment and analysis*

The detection model improvements in this paper include the two-branch backbone feature extraction network (BFFS), the spatial attention module (ASAM), and the Focal-CIO. To explore which improvement is more important to improve the performance, we conducted ablation experiments on Seaships datasets. We can see the results of the experiments in Table 1. Experiment A is replacing the loss function, and the performance of the detection model can be found when Focal-CIOU is used as the loss function. It can be seen that the new loss function improves the mAP50-95 by 0.4%, which also indicates that the loss function is more suitable for this dataset. Experiment B integrates ASAM into YOLOX. It can be found that the introduction of the ASAM module improves mAP50-95, Recall, and Precision by 0.75%, 2.33%, and 2.35%, respectively. This indicates that the ASAM module achieves better perception capability on multi-scale fishing vessel targets. Experiment C uses a new backbone network, and after introducing the dual architecture with the BFFS strategy, it makes the mAP50-95, Recall, and Precision improve by 1.46%, 2.99%, and 3.43%, respectively. This indicates that the dual-branch network enriches the fine-grained features of fishing vessels, enhancing the network's ability to recognize local features in images. And for our ASDNet algorithm, mAP50-95 reaches 0.8025.

**Table 1.** Results of ablation experiments on the Fishing-ours.

| Models | ASAM | BFFS | Focal-CIOU | P | R | Params | mAP50-95 |
|---|---|---|---|---|---|---|---|
| YOLOX | × | × | × | 0.9543 | 0.9516 | 8.94 | 0.7786 |
| A | × | × | √ | 0.9527 | 0.9653 | 8.94 | 0.7826 |
| B | √ | × | × | 0.9778 | 0.9749 | 8.95 | 0.7861 |
| C | × | √ | × | 0.9886 | 0.9815 | 9.23 | 0.7932 |
| D | √ | √ | × | 0.9935 | 0.9871 | 9.25 | 0.8013 |
| E(ASDNet) | √ | √ | √ | 0.9964 | 0.9921 | 9.25 | **0.8025** |

### *3.4 Detection result with ASDNet*

To validate the model's performance, we analyzed its detection results on 300 test images taken in clear weather conditions. In the test set, there were a total of 576 fishing vessel targets. The model's detection results are shown in Table 2. The overall accuracy and recall of our model are 99.35% and 98.71%, respectively.

**Table 2.** Detection results for multi-scale fishing ships.

| Fishing | Number | Precision(%) | Recall(%) |
|---|---|---|---|
| Large scale | 194 | 95.43 | 95.16 |
| Middle scale | 195 | 97.78 | 97.49 |
| Small scale | 187 | 98.86 | 98.15 |
| Total | 576 | 99.35 | 98.71 |

In conclusion, our model demonstrates good detection performance when it comes to detecting fishing vessel targets at sea. Among them, Figure 5 displays real-world test cases of our model at sea. Our model has achieved excellent results, demonstrating strong adaptability and accurate identification of fishing vessels across various scales—large, medium, and small. Overall, the performance aligns with our expectations.

**Figure 5.** Our case of monitoring fishing vessels in harbor areas using the ASDNet algorithm.

Journal of Physics: Conference Series **2813** (2024) 012001 doi:10.1088/1742-6596/2813/1/012001

## 4. Conclusion

Aiming at the large misdetection and omission problem when detecting fishing vessel targets in ports, this paper proposes ASDNet with YOLOX as the baseline: firstly, integrating adaptive spatial attention modules into the backbone network enhances the network's perception capability of multi-scale fishing vessel targets; secondly, the dual-branch backbone network is introduced, and the dual-branch feature fusion is performed by using bilateral feature enhancement and fusion strategy to solve the problem of fishing vessel target profile and feature loss, and finally, the two-branch backbone network is introduced. In the Fishing-ours dataset, mAP50-95 is improved by 2.39%, which proves the network effectiveness of ASDNet and achieves better results in actual detection.

## References


[1] Zhou, S., & Yin, J. (2022, January). YOLO-Ship: A Visible Light Ship Detection Method. In 2022 2nd International Conference on Consumer Electronics and Computer Engineering (ICCECE) (pp. 113-118). IEEE.

[2] Zhao, H., Zhang, H., & Zhao, Y. (2023). Yolov7-sea: Object detection of maritime UAV images based on improved yolov7. In Proceedings of the IEEE/CVF winter conference on applications of computer vision (pp. 233-238).

[3] He, G., Wang, W., Shi, B., Liu, S., Xiang, H., & Wang, X. (2022). An improved yolov4 algorithm-based object detection method for maritime vessels. International Journal of Science and Engineering Applications, 11, 50-55.

[4] Zhao, K., Lu, R., Wang, S., Yang, X., Li, Q., & Fan, J. (2023). ST-YOLOA: a Swin-transformer-based YOLO model with an attention mechanism for SAR ship detection under complex background. Frontiers in Neurorobotics, 17, 1170163.

[5] Li, Y., Yuan, H., Wang, Y., & Xiao, C. (2022). GGT-YOLO: a novel object detection algorithm for drone-based maritime cruising. Drones, 6(11), 335.

[6] Hu, J., Zhi, X., Shi, T., Zhang, W., Cui, Y., & Zhao, S. (2021). PAG-YOLO: A portable attention-guided YOLO network for small ship detection. Remote Sensing, 13(16), 3059.

[7] Sinha, D., & El-Sharkawy, M. (2019, October). Thin Mobilenet: An enhanced Mobilenet architecture. In 2019 IEEE 10th Annual Ubiquitous Computing, Electronics & Mobile Communication Conference (UEMCON) (pp. 0280-0285). IEEE.

[8] Wang, C. Y., Liao, H. Y. M., Wu, Y. H., Chen, P. Y., Hsieh, J. W., & Yeh, I. H. (2020). CSPNet: A new backbone that can enhance the learning capability of CNN. In Proceedings of the IEEE/CVF conference on computer vision and pattern recognition workshops (pp. 390-391).

[9] Woo, S., Park, J., Lee, J. Y., & Kweon, I. S. (2018). Cbam: Convolutional block attention module. In Proceedings of the European Conference on Computer Vision (ECCV) (pp. 3-19).

[10] Du, S., Zhang, B., Zhang, P., & Xiang, P. (2021, July). An improved bounding box regression loss function based on CIOU loss for multi-scale object detection. In 2021 IEEE 2nd International Conference on Pattern Recognition and Machine Learning (PRML) (pp. 92-98). IEEE.

[11] Lin, T. Y., Goyal, P., Girshick, R., He, K., & Dollár, P. (2017). Focal loss for dense object detection. In Proceedings of the IEEE International Conference on Computer Vision (pp. 2980-2988).

# Design of diesel engine vibration signal acquisition system based on Zynq

**Huishuai Wang[1], Haibo Gao[1,*] and Wen Zuo[2]**

[1]School of Naval Architecture, Ocean and Energy Power Engineering, Wuhan University of Technology, Wuhan 430063, China
[2]Wuhan Hailing Optoelectronic Technology Co., LTD, Wuhan 430074, China

*Corresponding author's e-mail: hbgao@whut.edu.cn

**Abstract.** The vibration signal of a diesel engine contains a large amount of fault information, and an accurate collection of vibration signals will provide reliable data support for diesel engine condition monitoring. This paper designs a diesel engine vibration signal acquisition system based on the powerful Zynq chip. IEPE piezoelectric acceleration sensors are selected to collect diesel engine vibration signals, and a signal conditioning circuit is designed to filter out interference during signal transmission using the characteristics of differential signals and to adjust the voltage amplitude of the signal to ensure compliance with the input voltage level requirements of the AD conversion module. Considering the excellent performance of the AD7771, it is selected as the AD conversion chip, and the peripheral circuit design is completed. With the help of Vivado software, the software and hardware design based on the Zynq chip is completed, controlling the AD7771 to sample 8 channels of vibration signals. Finally, a signal conditioning circuit simulation model is established in Multisim 14.1 software to verify its good performance, and upper computer software is developed to monitor the fault information of the diesel engine.

## 1. Introduction

Vibration is a common physical quantity in the field of mechanical engineering. Once a diesel engine enters the operating state, it is accompanied by vibration[1]. The vibration signal of a diesel engine contains a large amount of information that can characterize the working state of the engine, including fault information[2]. Therefore, collecting the vibration signal of a diesel engine can be used for fault diagnosis. Currently, the vibration signal acquisition systems for diesel engines on the market ensure good anti-interference capability by separately making the signal conditioning module into a conditioning board[3]. Although this ensures good performance, it poses many inconveniences in practical applications and increases the cost of the system[4]. In addition, the core controller of the current vibration signal acquisition system typically only uses FPGA or ARM[5], with FPGA having weak scheduling capabilities[6] and ARM processing data at a slower speed[7], neither of which can meet the requirements of this paper.

In 2010, Xilinx introduced the Zynq-7000 processing platform, which combines FPGA with dual-core ARM Cortex-A9 processors. This integration not only combines the performance of both, but also achieves low power consumption, standardization, and a dedicated data acquisition and processing solution. At the same time, this platform uses the AXI high-speed bus as a bridge between FPGA and ARM, allowing the two to closely integrate into a whole, fully leveraging their respective advantages to

Journal of Physics: Conference Series **2813** (2024) 012002 doi:10.1088/1742-6596/2813/1/012002

work together[8]. To achieve state monitoring and fault diagnosis of diesel engines through collecting vibration signals, this paper selects Zynq as the main control chip, making use of the FPGA's high-speed parallel computing capabilities and ARM's real-time analysis and control capabilities to design a diesel engine vibration signal acquisition system[9]. This system will collect real-time vibration signals of the diesel engine, analyze the data, and monitor the health status of the diesel engine[10]. The system framework is shown in Figure 1.

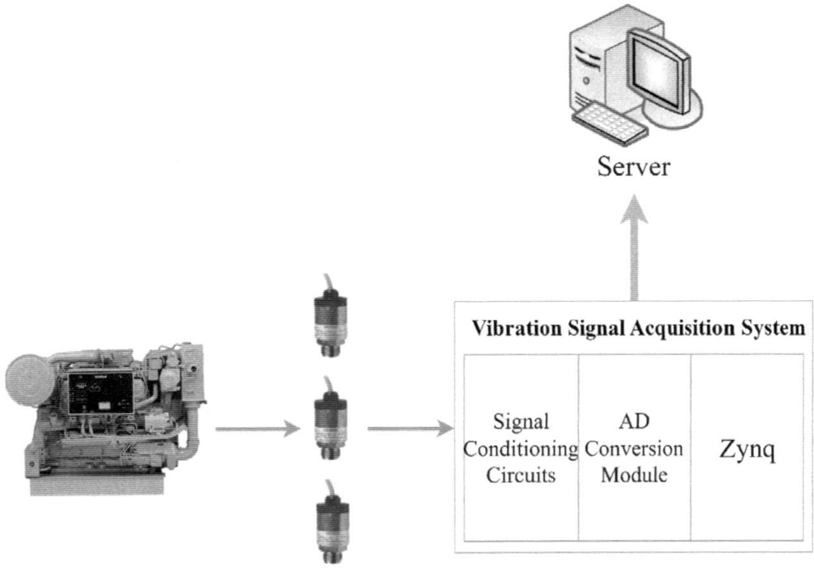


**Figure 1.** Diesel engine vibration signal acquisition system architecture.

## 2. Selection of vibration measurement sensor

According to the vibration severity measurement frequency range specified in the GB/T 7184-2023 standard for reciprocating internal combustion engine vibration assessment, it is known that the frequency range of conventional vibration generated by diesel engines is 2Hz-1kHz. The vibration frequency of diesel engines is not high, and the vibration acceleration signals in the low-frequency range are weak. High-impedance charge signals are very susceptible to interference. Considering the characteristics of acceleration sensors, IEPE piezoelectric acceleration sensors are more suitable for measuring diesel engine vibration signals[11]. This design measures eight channels of 2Hz-1kHz vibration signals using the Huishi technology's IEPE single-axis piezoelectric acceleration sensor CA3010A-10. The measurement range and sensitivity can meet the requirements of measuring diesel engine vibration signals, and the sensor is lightweight with a small input impedance. The main parameters of this sensor are shown in Table 1.

**Table 1.** CA3010A-10 main parameters.

| Project | Parameter |
| --- | --- |
| Sensitivity /mV/g | 10 |
| Dynamic range /g | ±500 |
| Frequency response /kHz | 1~10 |
| Output impedance /Ω | <100 |
| Full-scale output /V | ±5 |
| Operating temperature /°C | -55~125 |

## 3. Signal conditioning circuit design

The vibration measurement sensor selected is an IEPE piezoelectric accelerometer sensor, which is powered by a constant current source. The sampled data output includes a DC component and weak low-frequency AC signals. The AC signal truly reflects the vibration state of the diesel engine. The main function of the signal conditioning circuit is to provide a constant current source to the sensor, isolate the DC component in the output signal, and filter out noise interference[12].

The constant current source power circuit is shown in Figure 2. The XTR111 is a precision voltage current converter designed for standard 0mA-20mA or 4mA-20mA Analog signals, supporting up to 36mA current output. This chip is powered by a 24V DC power supply with an input voltage of 2.5V. The ratio between the input voltage and output current is set by an external resistor connected to the SET pin, with the calculation formula as follows:

$$I_{OUT} = 10 \times \frac{VIN}{R225+R222} \tag{1}$$

In the formula, $I_{OUT}$ represents the output current, and $VIN$ represents the input voltage. By setting the resistance values of $R225$ and $R222$ appropriately, a constant current output signal of 4mA can be obtained as the excitation current for the sensor.

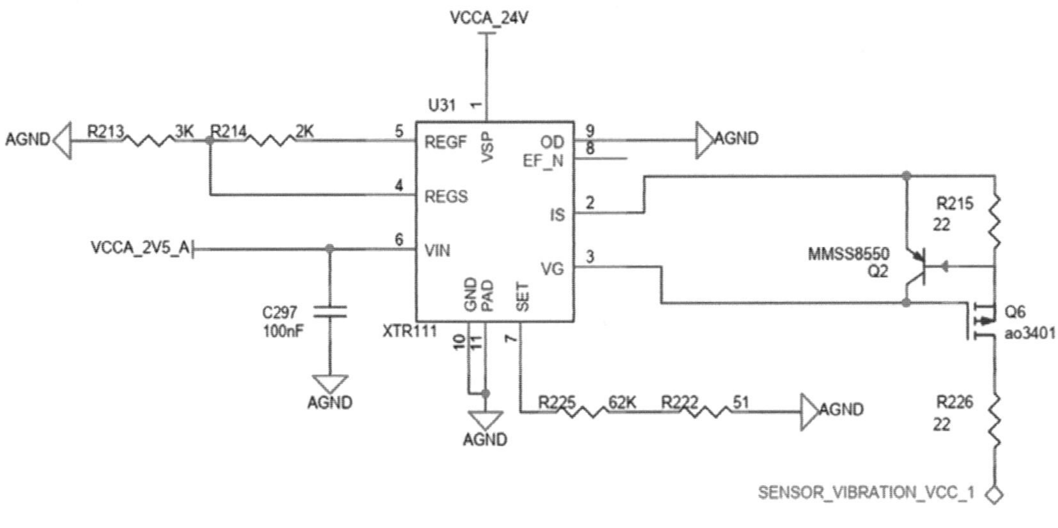


**Figure 2.** Constant current source power supply circuit.

The vibration signal is transmitted in a complex electromagnetic environment, with various signal interferences present in the surroundings. It is inevitable that impurity signals may be introduced into the sampled data, resulting in abnormal sampled data. An IEPE piezoelectric accelerometer sensor is connected by two signal lines: one is used to power the sensor and is connected to the constant current power supply circuit, and the other is the sensor's signal output end, transmitting the voltage signal reflecting the real vibration state of the diesel engine. Due to the relatively weak voltage signal output by the sensor, interference signals can easily be introduced from the sensor to the data acquisition system, necessitating a signal conditioning circuit to eliminate these interferences.

The circuit shown in Figure 3 is an input signal subtraction circuit implemented using an operational amplifier OPA277UA and external resistors and capacitors. OPA277UA is a powerful operational amplifier that provides improved noise performance, a wider output voltage swing, and double the speed compared to its predecessor, with only half the static current. OPA277UA is powered by ±12V voltage, with power supply input pins VCCA_12V and VEEA_-12V.

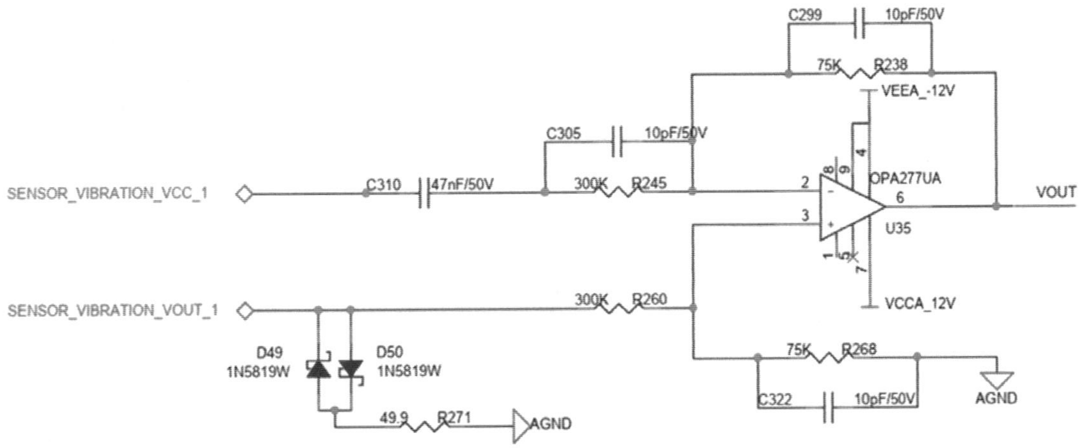


**Figure 3.** Input signal subtraction circuit.

The signal output terminal SENSOR_VIBRATION_VOUT_1 of the IEPE piezoelectric accelerometer sensor is connected to the positive input terminal of the OPA277UA. Protection against excessive input voltage is provided in the signal connection path through series resistors and diode clamps, capable of withstanding ±30V input voltage without damage. When the input exceeds a certain threshold, the protection diodes will conduct current. The sensor constant current power supply terminal SENSOR_VIBRATION_VCC_1 is connected to the negative input terminal of the OPA277UA. Capacitor C310 is used to isolate the DC component of the constant current source circuit output, while the capacitor C299, in parallel with the feedback resistor R238, is used to prevent operational amplifier self-oscillation from suppressing high-frequency noise. Additionally, capacitors C305 and C322 are also used to suppress high-frequency noise. The two signal lines of the sensor are twisted pair cables, with interference signals acting on both lines simultaneously. By utilizing a subtraction circuit, interference signals can be filtered out. The output voltage is obtained by subtracting the two input signals, with the calculation formula as follows:

$$VOUT = (V3 - V2) \times \frac{R238}{R245} \tag{2}$$

In the formula, $V3$ and $V2$ represent the voltage inputs to the positive and negative terminals of the operational amplifier, respectively. $VOUT$ is the result of the circuit operation, indicating the output voltage after filtering out interference signals. The values of $R238$ and $R245$ are chosen to adjust the magnitude of *VOUT*, where *VOUT's* magnitude is reduced by a factor of four.

To meet the sampling requirements of the ADC and enhance signal immunity to interference, the single-ended signal output by the OPA277UA needs to be converted to a differential signal. The circuit shown in Figure 4 achieves this by using the OPA2277UA. The OPA2277UA features two sets of input channels, enabling it to function as an inverter and a voltage follower. It converts the input signal VOUT into differential signals ADCURRENT_A_P and ADCURRENT_A_N, which are then connected to the ADC module. VCM, provided by the ADC, offers a DC bias for signal transmission, balancing the operation of the ADC and meeting the requirements of input voltage levels.

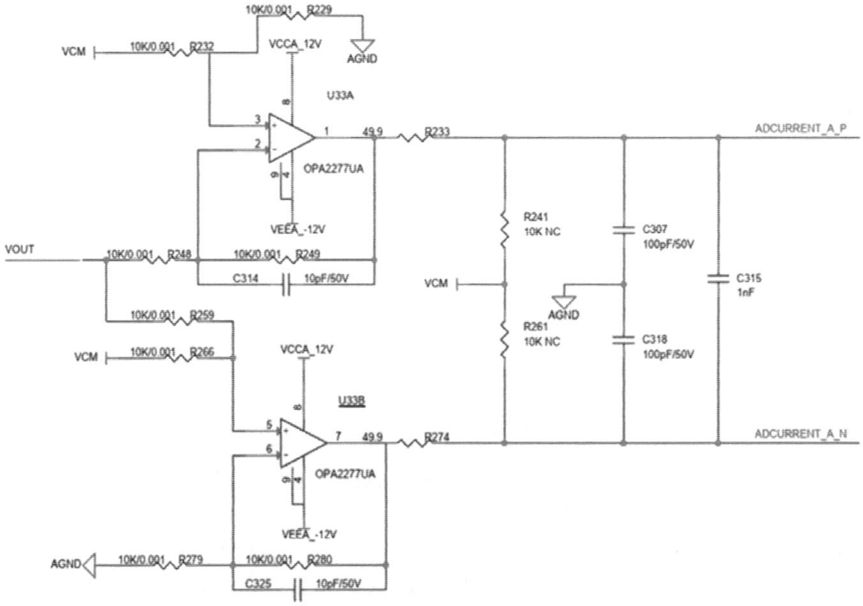

**Figure 4.** Single-ended to differential signal conversion circuit.

## 4. AD conversion module design

### *4.1. Sampling theorem*

ADC conversion involves sampling the Analog signal output from the signal conditioning circuit at a certain sampling frequency to represent the Analog signal using discrete digital signals. The sampling frequency can be high or low, but it must be sufficient to recover the original Analog signal from the sampled signal. According to the Nyquist sampling theorem:

$$f_s \geq 2f_{imax} \tag{3}$$

In the formula, $f_s$ is the sampling frequency, and $f_{imax}$ is the highest frequency component of the analog signal. Increasing the sampling frequency requires the conversion circuit to operate at a faster speed, so the sampling rate cannot be increased indefinitely. Typically, $f_s$ is chosen to be 3 to 5 times $f_{imax}$.

### *4.2. ADC chip selection*

A synchronous sampling of eight channels of vibration signals at a sampling rate of 12.8 kps is required. Considering the system's accuracy and conversion speed requirements, the ANALOG DEVICES AD7771 Analog-to-Digital conversion chip is selected. This chip is widely used in applications such as power quality measurement, general data acquisition, and industrial process control. The AD7771 is an 8-channel synchronous sampling ADC chip, internally employing 8 Σ-Δ ADC conversions. The chip provides ultra-low input current, allowing direct sensor connection. Each input channel has a programmable gain stage with gains of 1, 2, 4, and 8, mapping low-amplitude sensor outputs to the full-scale ADC input range, thereby providing a large dynamic range for the signal chain. The AD7771 accepts VREF from 1V to 3.6V, with Analog inputs accepting unipolar or true bipolar Analog input signals, and Analog power voltages being unipolar 3.3V or bipolar ±1.65V. The Analog input can be configured to accept true differential or single-ended signals to match different sensor output configurations. Other technical specifications include:

- Resolution: 24 bits
- Digital I/O Power: 1.8V to 3.6V

Journal of Physics: Conference Series **2813** (2024) 012002 doi:10.1088/1742-6596/2813/1/012002

- Bipolar Input Range: ±1.8V
- Output Data Rate: Up to 128ksps per channel
- Signal-to-Noise Ratio (SNR): 95dB
- Total Harmonic Distortion (THD): -109dB
- Operating Temperature Range: -40°C to 125°C

The AD7771 offers high-resolution and low-power modes for selection. In high-resolution mode, with a main clock frequency of 8192 kHz, the maximum output sampling rate is 128 ksps, with power consumption ranging from 133 to 152 mW. In low-power mode, with a main clock frequency of 4096 kHz, the maximum output sampling rate is 32 ksps, with power consumption ranging from 42 to 48.5 mW.

### *4.3. ADC Circuit Design*

After selecting the ADC chip, peripheral circuit configuration is necessary to ensure proper operation. Circuit configuration includes power supply, input-output signal connections, and control signal connections. The external interface connection circuit for the AD7771 is shown in Figure 5.

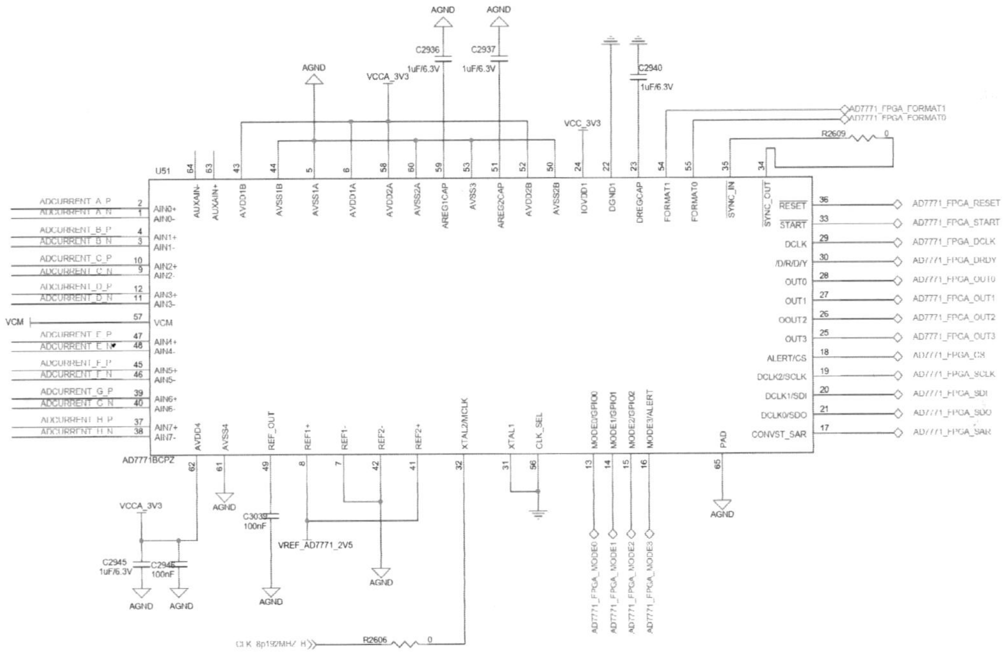


**Figure 5.** AD7771 external interface connection circuit.

The power supply includes Analog power, digital power, and internal reference voltage sources. The analog power uses a single-ended 3.3V DC supply connected to analog ground; the digital power uses a 3.3V DC supply connected to analog ground; and the internal reference voltage source is 2.5V, which is connected to analog ground. The VCM pin provides a buffered common-mode voltage output, which can bias the Analog input signal. The potential of VCM satisfies the following formula:

$$VCM = (AVDD1A - AVSS1A)/2 \tag{4}$$

In the formula, *AVDD1A* is the Analog voltage input pin, and *AVSS1A* is the Analog ground input pin. Since the chip uses a single-ended 3.3V DC power supply, the potential of VCM is 1.65V.

The 8-channel vibration signal conditioning is converted to differential signals and connected to AIN0~AIN8 of the ADC chip. This chip supports differential, pseudo-differential, and single-ended inputs. Differential inputs have the advantages of strong anti-interference capability, effective suppression of EMI, and precise timing positioning, so differential inputs are chosen. The AD7771 provides two interfaces for data output and SPI control, with the interface selection controlled through the FORMAT0 and FORMAT1 pins. The data output interface is connected directly to the FPGA via hard lines, enabling different sampling functions through pin configurations. The SPI control interface is more flexible, allowing control through configuring relevant registers with more comprehensive functions, but it is not as fast as the data output interface. The pins related to control and output are directly connected to the FPGA, which is responsible for controlling and receiving the sampled data for subsequent data processing.

## 5. Development process of Zynq

In this design, the main control chip selected is the Xilinx Zynq-7000 series SoC, with the model number XC7Z020CLG484. It is divided into the PS (Processing System) side and the PL (Programmable Logic) side, interconnected internally through the AXI bus. The system diagram of the Zynq chip is shown in Figure 6. The PL side features 85,000 logic cells, 4.9Mbits of embedded memory resources, 220 DSP units, 4 clock management units, 16 global clock networks, 6 user IO banks, and a maximum of 253 user IOs. The PS part integrates two Cortex-A9 processors, internal memory, external memory, interfaces, and peripheral controllers. The main parameters of the PS side are as follows:

- Application processor based on ARM dual-core Cortex-A9, operating on ARM-v7 architecture, with a core frequency of up to 766MHz
- Integrated on-chip Boot ROM and 256KB on-chip RAM
- Integrated external memory interfaces, supporting 16/32-bit DDR2 and DDR3 interfaces
- Integrated Gigabit Ethernet controller: supporting GMII, RGMII, and SGMII interfaces
- Supports two USB 2.0 OTG interfaces, each supporting up to 12 nodes
- Supports two CAN 2.0B bus interfaces
- Includes two SD card, SDIO, MMC compatible controllers
- Includes 2 SPI, 2 UART, and 2 I2C controllers
- 4 sets of 32-bit GPIO, 54 MIOs as PS system IOs, and 64 EMIOs connected to PL

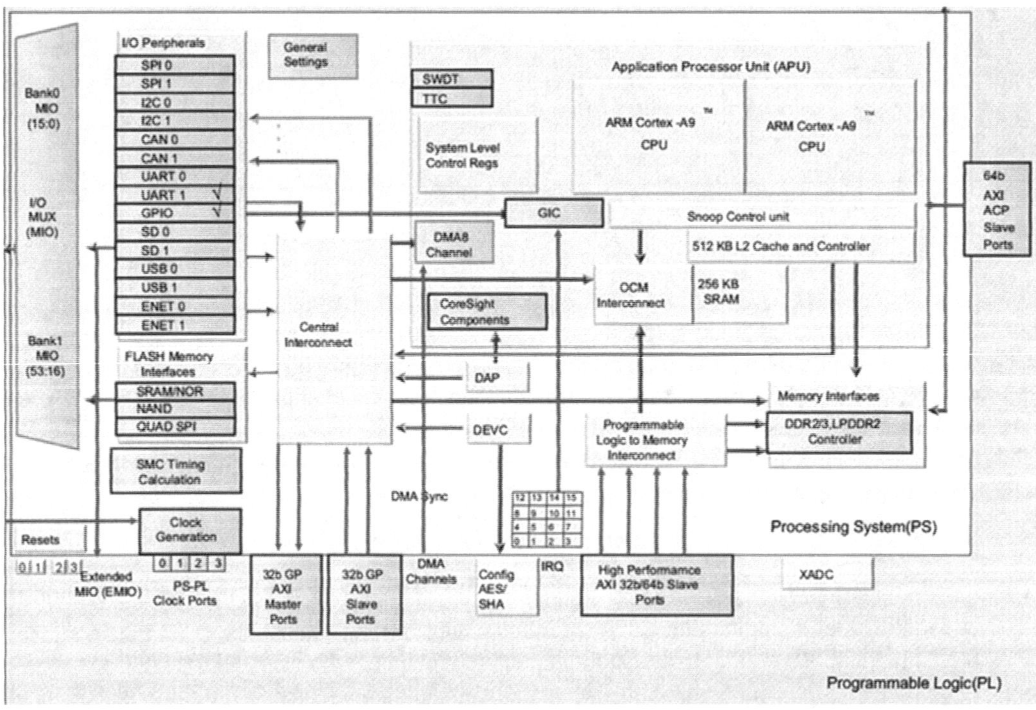


**Figure 6.** Zynq chip system block diagram.

Zynq development involves the development of both the PL and PS sides, which are implemented through the Xilinx Vivado integrated development environment. The basic development process includes hardware design and software design[13]. Firstly, a new project is created using the Vivado software, and then the processor system is created using the IP integrator. The processor's peripheral circuits and device connections are completed according to requirements. On the PL side, logic programs are written in Verilog to control the AD7771 for sampling, data processing, and transmission. Finally, the top-level HDL is generated, concluding the hardware design phase. The software design involves importing the generated HDL into the Vivado SDK software. In the SDK, C/C++ is used to develop application programs to implement data storage and communication functions.

## 6. System testing

### *6.1. Signal conditioning circuit testing*

IEPE piezoelectric accelerometer sensors are installed on diesel engines. The vibration signals collected from the sensors go through a complex working environment before reaching the signal acquisition system. The transmitted voltage signal has a small amplitude and is susceptible to interference. Therefore, the quality of the signal conditioning circuit is crucial for the normal operation of the system. The signal conditioning circuit processes the Analog voltage signal output by the sensor. A simulation model is built using Multisim 14.1 software to verify the basic functionality of the signal processing circuit. The simulation model of the signal conditioning circuit is shown in Figure 7.

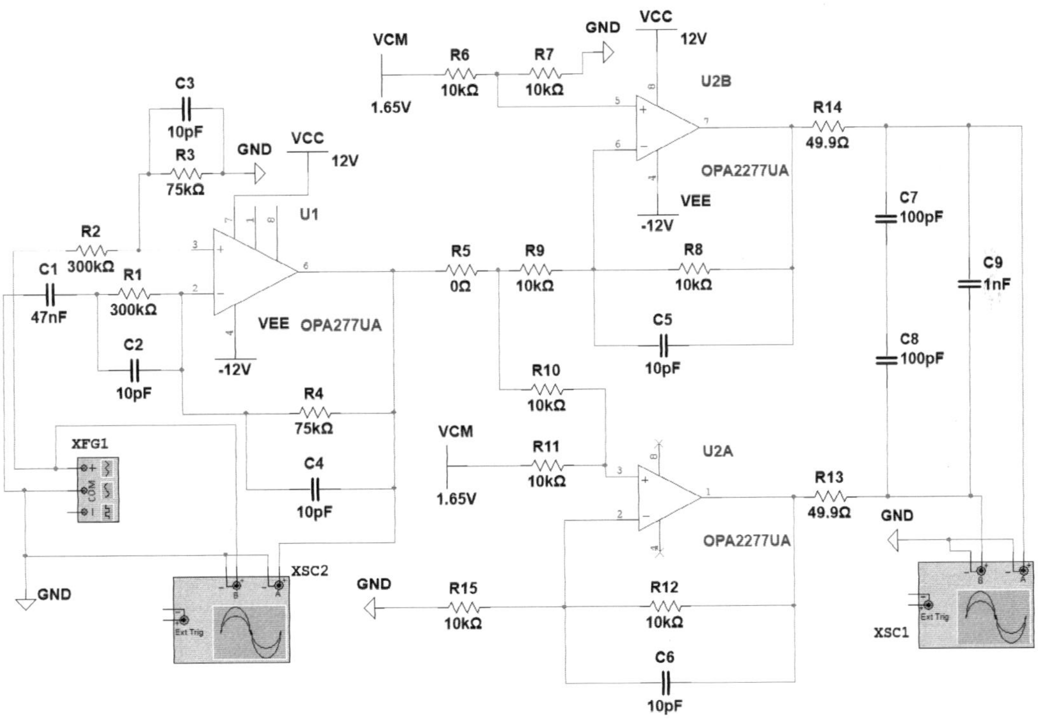


**Figure 7.** Signal conditioning circuit simulation model.

In Figure 7, XFG1 is a function generator that provides a 5V amplitude, 100Hz frequency sine signal, simulating the output of the IEPE accelerometer sensor. XSC1 and XSC2 are two oscilloscopes, with channels A and B of XSC1 connected to the output of the signal conditioning circuit, channel A of XSC2 connected to the output of the subtraction circuit, and channel B connected to the function generator. After starting the simulation, wait for the system to stabilize, and the simulation results are shown in Figures 8 and 9.

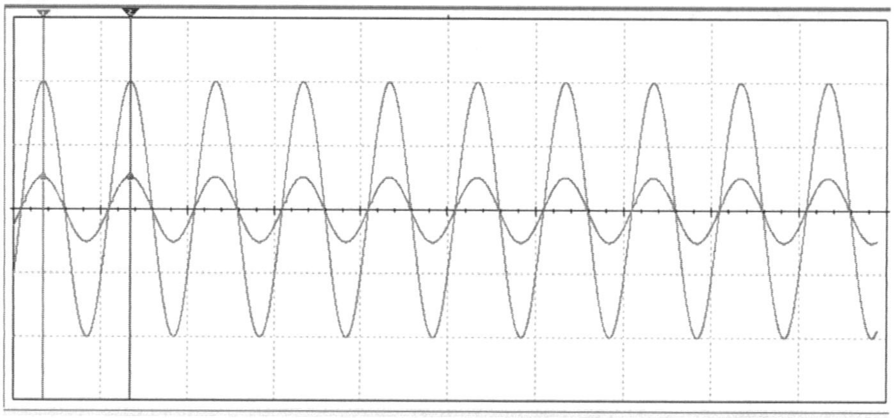

**Figure 8.** XSC2 display value.

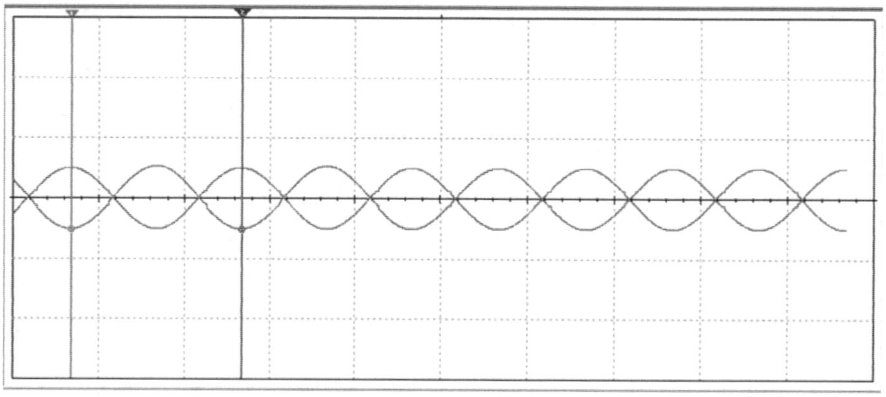

**Figure 9.** XSC1 display value.

In Figure 8, the Analog voltage output from the function generator is reduced by a factor of four after passing through the subtraction circuit, with the frequency remaining unchanged. The simulation results are consistent with the design requirements. Figure 9 shows the differential signal output from the signal conditioning circuit, which serves as the input to the AD conversion module. The signals on channels A and B are equal in magnitude but opposite in direction. The difference between the two signals represents the actual value, and the simulation results align with the design requirements. Through simulation testing, the excellent performance of the signal conditioning circuit designed in this paper has been verified.

### *6.2. The upper computer software*

The data source for the upper computer software is the original vibration signals and their characteristics obtained from the diesel engine vibration signal acquisition system. After processing, this data is used to diagnose the fault of the diesel engine. The upper computer software interface can display fault diagnosis records. In the event of abnormal vibration faults in the diesel engine, the diagnosis records show details such as fault time, recovery time, fault severity, and auxiliary decision-making content. This setup effectively monitors the health status of the diesel engine, providing support for timely fault detection and resolution.

## 7. Conclusion

This paper presents the design of a diesel engine vibration signal acquisition system, which involves analyzing and processing the collected vibration signals for use in diesel engine fault diagnosis. Due to the low frequency of diesel engine vibration signals and the weak output voltage signal from the IEPE piezoelectric accelerometer sensor, which is susceptible to interference during transmission, a signal conditioning circuit was designed to eliminate interference and adjust the voltage amplitude. A simulation model of the signal conditioning circuit was built using Multisim 14.1 software, with oscilloscope connections to measure the output of the subtraction circuit and signal conditioning circuit. The results indicate that the signal conditioning circuit designed in this paper effectively filters out interference signals and adjusts the voltage to an appropriate level. To ensure sampling accuracy, the AD7771 chip with a resolution of 24 bits was selected, and peripheral circuits were designed. The signal conditioning circuit, AD conversion module, and Zynq chip were integrated into a single acquisition board, enhancing usability and cost savings. An upper computer software was developed, with the fault diagnosis record interface displaying alarm records caused by abnormal vibrations, demonstrating the normal operation of the vibration signal acquisition system and providing data support for the upper computer software.

**Acknowledgments**
Thanks to the support and assistance from the Electric Propulsion Laboratory of Wuhan University of Technology and Wuhan Hailing Optoelectronic Technology Co., Ltd., the experimental equipment and research funding used in this paper would not have been possible without their help.

**References**
[1] Taghizadeh-Alisaraei A, Mandavian A. Fault detection of injectors in diesel engines using vibration time-frequency analysis[J]. Applied Acoustics, 2019, 143(JAN.):48-58.
[2] Ayati S M. Fault detection and diagnosis of a 12-cylinder trainset diesel engine based on vibration signature analysis and neural network[J]. Proceedings of the Institution of Mechanical Engineers, Part C. Journal of Mechanical Engineering Science, 2019, 233(6).
[3] ZHAO R, MA Y. Research and Design of Data Acquisition System Based on Zynq [J]. Electronic Testing, 2016(22):2.
[4] Zhiqiang Y, Bin H, Hong J, et al. Design of multi-channel synchronous real-time high-speed data acquisition system based on FPGA[J]. Application of Electronic Technique, 2019.
[5] Bengherbia B, Ould Zmirli M, Toubal A, et al. FPGA-based wireless sensor nodes for vibration monitoring systems and fault diagnosis[J]. Measurement, 2017, 101:81-92.
[6] Ouyang Y, Li J, Zhang B, et al. Real-time processing system for cavity ring down signal based on ZYNQ chip[J]. IEICE Electronics Express, 2022, 19(1):20210384-20210384.
[7] Perez-Castillo A J, Morales-Caporal R, Rangel-Magdaleno J D J, et al. Real Time Monitoring of 3 Axis Accelerometer using an FPGA Zynq - 7000 and Embedded Linux through Ethernet[C]//15th International Conference on Electrical Engineering, Computing Science and Automatic Control.2018.
[8] HE Bin, Xilinx All Programmable Zynq-7000 SoC Design Guide [M], Beijing: Tsinghua University Press, 2013.
[9] ZHANG X J, CHEN J Y. Design of High-Speed Data Acquisition System Based on Zynq [J]. Journal of East China Jiaotong University, 2020, 37(6):7.
[10] Tan J, He L, Wang J. Signal acquisition and processing system based on zynq dual core[J]. MATEC Web of Conferences, 2018, 246:03001.
[11] XU B F. A New High-Performance IEPE Acceleration Sensor [J]. Electronic Testing, 2021. 5.
[12] LI Z H, JIAO X Q, JIA X Z. Design of a Conditioning Circuit for Piezoelectric Acceleration Sensor [J]. Instrumentation Technology and Sensor, 2017(10):5.
[13] Yu M. Film and Television Culture Dissemination Based on ZYNQ Embedded Digital Image Processing [J]. Microprocessors and Microsystems, 2021, 82(9): 103921.

# A UWB indoor positioning method based on triple filtering fingerprint matching algorithm

**Jiaqi Yang*, Junbo Gao, Wei Sun and Xin Jing**

College of Information Engineering, Shanghai Maritime University, Shanghai, China

*Corresponding author's e-mail: 1415056681@qq.com

**Abstract.** In recent years, ultra-wideband (UWB) has gradually become a research hot spot in the field of indoor positioning because of its various advantages. Although UWB has such excellent performance in normal environments, the non-line-of-sight propagation of signals in complex indoor environments and the multi-path effect caused by obstacles will affect its positioning accuracy. To solve this problem, we use the fingerprint positioning method and optimize the previously commonly used k-nearest neighbor algorithm in the online matching phase. In this paper, we proposed a triple-filtered k-nearest neighbor algorithm based on sample distance weighting (TFWKNN). Experimental results show that the mean calculation error of TFWKNN is 63.4% less than the k-nearest algorithm, and the proposed algorithm has better prediction stability than other commonly used machine learning regression algorithms.

## 1. Introduction

In recent years, indoor positioning systems have received widespread attention due to their wide application in various fields such as medical, retail, and logistics. Among many indoor positioning technologies, ultra-wideband (UWB) has become the most promising candidate due to its unique characteristics and functions [1]. Unlike traditional positioning technologies such as Wi-Fi and Bluetooth, UWB offers superior accuracy, robustness to multi-path interference [2], and high immunity to environmental factors such as signal attenuation and reflections. At the same time, it has the characteristics of a wide spectrum, high speed, low power consumption, and so on [3].

Although UWB has many advantages, its traditional positioning algorithms are mostly implemented based on ranging principles, such as the TOA algorithm and TDOA algorithm [4]. However, non-visual detection in complex environments will cause the application of traditional algorithms in complex indoor environments to face some challenges. To solve this problem, some researchers optimize traditional positioning algorithms [5], while others hope to use classifiers to identify interfered signals and correct them [6]. We choose to use the fingerprint positioning method [7] and propose a new UWB fingerprint positioning online matching algorithm called TFWKNN.

## 2. TFWKNN algorithm

The TFKWKNN algorithm optimizes the original algorithm based on the problems encountered by WKNN in weight calculation in the fingerprint positioning task and the K value distribution of the test data under the minimum error condition. The algorithm flow chart is shown in Figure 1.

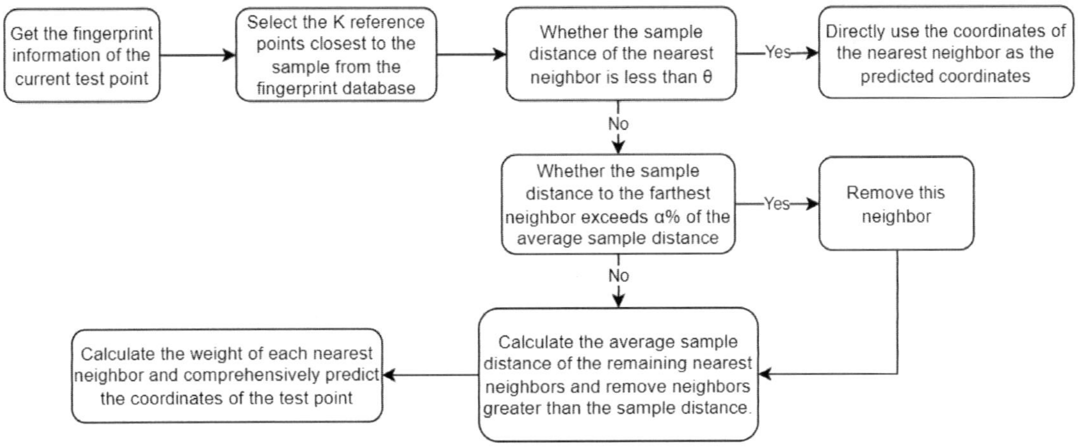


**Figure 1.** TFWKNN algorithm execution flow chart.

### *2.1. Optimization of weight calculation*

As a commonly used algorithm for fingerprint positioning online matching, the KNN algorithm has the advantages of simple implementation and fast calculation speed [8]. The implementation principle is shown in Formula (1). The calculation principle of this algorithm is to average the coordinates of the K nearest neighbors. Since it sets the calculation weights of all K nearest neighbors to be the same, and all are 1, in the indoor positioning environment, if the grid division of the fingerprint database data, those neighbors that are far away from the test point sample will cause the final positioning coordinates to deviate from the actual position.

$$(x, y) = \frac{\sum_{i=1}^{K}(x_i, y_i)}{K} \tag{1}$$

Due to the shortcomings of the KNN algorithm, the KNN algorithm based on sample distance weighting was later proposed. The weight calculation principle is shown in Formulas (2), (3), and (4). In indoor positioning tasks, the test point may happen to reach a certain reference point position in the fingerprint database during the actual application, so the $d_i$ will take zero, and its probability increases with the increase in mesh density in the indoor environment. To solve the "Zero-Division-Error: division by zero" caused by the division by zero that occurs during the calculation process and the inf element that appears when the array is divided by zero, we proposed to set a threshold to filter the nearest neighbor of the current sample before calculating the weight.

$$(x, y) = \frac{\sum_{i=1}^{K} w_i * (x_i, y_i)}{K} \tag{2}$$

$$w_i = \frac{\frac{1}{d_i}}{\sum_{i=1}^{K}\frac{1}{d_i}} \tag{3}$$

$$d_i = \sqrt{(a_i - a)^2 + \ldots + (n_i - n)^2} \tag{4}$$

Compared with traditional solutions, such as skipping calculation only when the sample distance is zero and adding a small number to the denominator, our method also improves the positioning accuracy in most scenarios compared with conventional solutions. The comparison results of these three methods are shown in Figure 2 and Figure 3.

Journal of Physics: Conference Series **2813** (2024) 012003 doi:10.1088/1742-6596/2813/1/012003

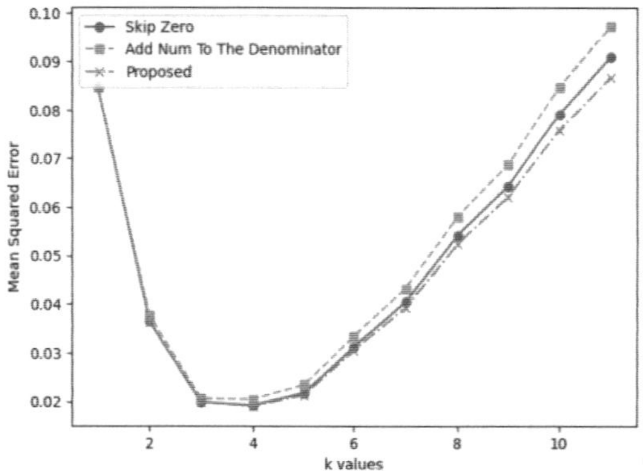


**Figure 2.** MSE of three methods under different K values.

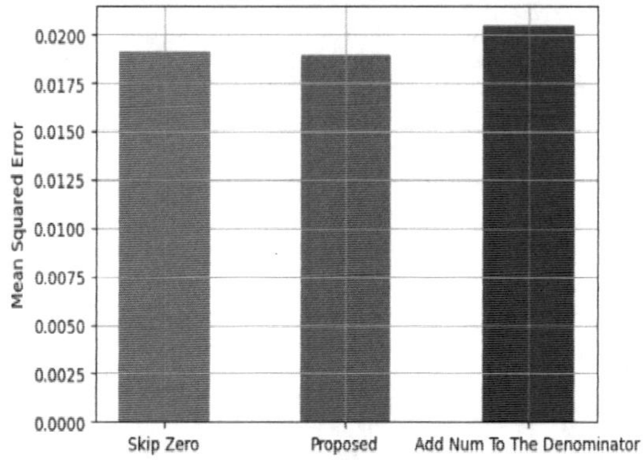


**Figure 3.** Minimum MSE comparison of three methods.

## *2.2. Triple filtering*

In the KNN algorithm, the selection of the K value has always been a problem. Because for different test data, the K value when achieving the best results is different in most cases. In this regard, I conducted corresponding experiments on the test data, predicted all the data in WKNN with different K values (1 to 15), and recorded all their errors. The experimental results of this experiment are shown in Figure 4. It can be seen from the figure that, in most cases, the smallest error is obtained when the K value is between 2 and 7. The proportion of data with the best-predicted coordinates under these six values is about 77.4%; K is 1, and 8 to 15 also have data that achieve the best prediction results, but these data account for a small amount.

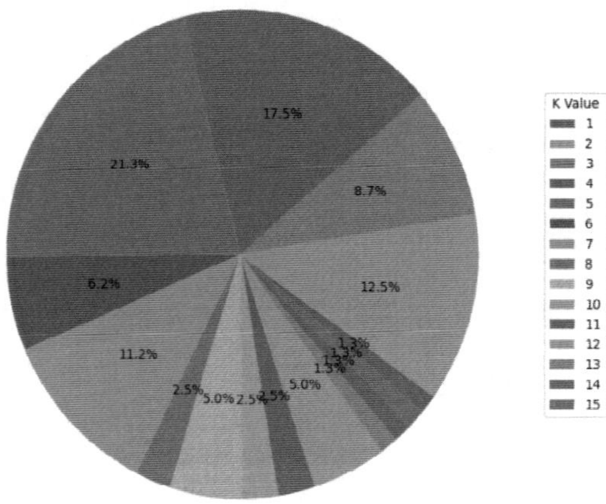


**Figure 4.** K value distribution when the test data obtains the minimum error.

Therefore, setting K to a fixed value and using it to predict positioning coordinates is not the best choice. If a dynamic number of neighbors is used in the prediction stage, the algorithm still has room for improvement. Based on the filtering in the previous section, I have proposed two filtering steps in the weighted calculation stage. The principle is to first obtain the reference point with the farthest sample distance among the K nearest neighbors before calculating the weight, and then obtain the distance from the current K nearest neighbor samples. The average values are compared, and if they exceed a certain proportion, they are excluded. Then, the mean value of the remaining nearest neighbors is calculated, and the nearest neighbor reference point that is less than or equal to the mean value is selected, and then weighted prediction is performed. The second filtering step is shown in

Formula (5). The last filtering is to take the average of the remaining neighbors and select the neighbors smaller than the distance for weighted calculation.

$$(d_{max} - d_{mean}) > \gamma * d_{mean} \tag{5}$$

## 3. Experiment

### 3.1. Experimental site

The collection of experimental data was carried out in the laboratory of the College of Information Engineering of Shanghai Maritime University. The entire experimental site is about 8 meters long and 5 meters wide. The field map and top view of the laboratory are shown in Figure 5 and Figure 6.

**Figure 5.** Experimental Site Photo.

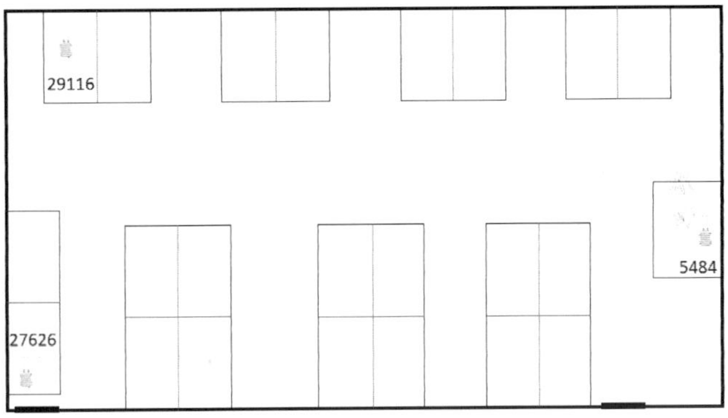


**Figure 6.** Experimental Site Plan.

The three numbers 5484, 29116, and 27626 in Figure 6 respectively represent the unique numbers of three UWB base stations. To construct the fingerprint database, the experimental site was also gridded in advance to determine the location distribution of all reference points.

### 3.2. Fingerprint database construction

The fingerprint positioning method is divided into two stages: offline and online [9]. The preparation work before the experiment is carried out in the offline stage. In the offline stage, we need to build a fingerprint database and select different data features as fingerprints based on different positioning technologies [10]. Since the original data cannot be used, it needs to be sorted to obtain the data format required by the fingerprint database data. After the fingerprint database is built, its internal data structure is shown in Table 1.

**Table 1.** Fingerprint database data structure.

| **Fingerprint Information** | | | | **Label** | |
|---|---|---|---|---|---|
| $feature_{11}$ | $feature_{12}$ | ... | $feature_{1n}$ | $x_1$ | $y_1$ |
| $feature_{21}$ | $feature_{22}$ | ... | $feature_{2n}$ | $x_2$ | $y_2$ |
| ... | ... | ... | ... | ... | ... |
| $feature_{m1}$ | $feature_{m2}$ | ... | $feature_{mn}$ | $x_m$ | $y_m$ |

At the same time, to ensure the quality of the reference point data in the fingerprint database, each grid position needs to be measured multiple times, and the reference point data collected over a continuous period. If there are outliers, they need to be deleted, and then the data are averaged, and finally integrate features and labels of all reference point data. This completes the fingerprint library construction work in the offline stage. The characteristics of the original data collected through experiments are shown in Table 2.

**Table 2.** Original data attribute table.

| Attribute name | Description |
|---|---|
| Time | Indicates data collection time |
| Rolling code | Encoding used to synchronize data |
| Anchor code | The unique identifier of the UWB base station |
| Tag code | The unique identifier of the UWB tag |
| Distance | Ranging value converted by TOF algorithm |
| RXL | Received signal strength |
| FPL | First received signal strength |
| Battery Voltage | Indicates the voltage intensity during data collection |

### *3.3. Experimental results*

In the test data, we selected a rectangular path to visually compare the performance of the TFWKNN algorithm. The results are shown in Figure 7. As can be seen in the figure, the predicted path of TFWKNN is closer to the real path and has better prediction stability.

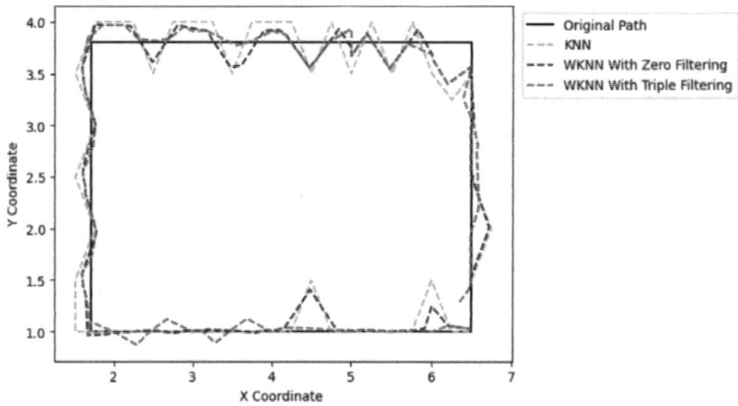


**Figure 7.** Comparison between TFWKNN and other online matching algorithms.

The comparison results between TFWKNN and other regression algorithms in machine learning are shown in Table 3. Experimental results prove that in the task of predicting coordinates for indoor positioning based on UWB, the minimum error, maximum error, and average error in the test data of TFWKNN are significantly lower than other machine learning algorithms.

**Table 3.** Comparison between TFWKNN and other machine learning algorithms.

| Algorithm | Minimum Error | Maximum Error | Mean Error |
|---|---|---|---|
| Linear Regression | 0.00011661 | 1.3608419 | 0.1432095 |
| Decision Tree | 0.00012094 | 2.6782345 | 0.1386814 |
| Random Forest | 0.00002527 | 2.1780795 | 0.0450333 |
| XGBoost | 0.00016747 | 1.9411895 | 0.0476716 |
| SVM | 0.00075639 | 1.0704303 | 0.1415946 |
| KNN | 0.00012335 | 0.8542989 | 0.0438529 |
| WKNN | 0.00001604 | 0.7164126 | 0.0191598 |
| **TFWKNN (Proposed)** | **0.00000099** | **0.5516332** | **0.0160457** |

## 4. Conclusions

Based on the problems encountered by UWB in indoor complex environment positioning, we proposed the fingerprint database online matching algorithm of TFWKNN. Compared with the KNN algorithm, it considers the weight relationship of the nearest neighbor reference point sample distance for the final prediction and solves the problem. Eliminating the anomaly of zero divisors when calculating weights also improves its accuracy in most cases. Afterward, considering the K value statistics of the test data when it reaches the minimum MSE in the weighted KNN algorithm, a dynamic number of nearest neighbors is used for the weighted prediction of coordinates through two filtering methods, which further improves the accuracy of the algorithm and at the same time in K When the value is large, the algorithm still maintains stable accuracy, which is better than the conventional online matching algorithm for fingerprint positioning.

**References**

[1] P. S. Farahsari, A. Farahzadi, J. Rezazadeh, and A. Bagheri, "A Survey on Indoor Positioning Systems for IoT-Based Applications," in IEEE Internet of Things Journal, vol. 9, no. 10, pp. 7680-7699, 15 May 15, 2022, DOI: 10.1109/JIOT.2022.3149048.

[2] R. J. Fontana, "Recent system applications of short-pulse ultra-wideband (UWB) technology," in IEEE Transactions on Microwave Theory and Techniques, vol. 52, no. 9, pp. 2087-2104, Sept. 2004, DOI: 10.1109/TMTT.2004.834186.

[3] M. Z. Win and R. A. Scholtz, "Impulse radio: how it works," in IEEE Communications Letters, vol. 2, no. 2, pp. 36-38, Feb. 1998, DOI: 10.1109/4234.660796

[4] S. Gezici et al., "Localization via ultra-wideband radios: a look at positioning aspects for future sensor networks," in IEEE Signal Processing Magazine, vol. 22, no. 4, pp. 70-84, July 2005, DOI: 10.1109/MSP.2005.1458289.

[5] L. Zhou and Y. Li, "A Novel UWB indoor Localization Algorithm based on TDOA in LOS/NLOS Environment," 2023 4th Information Communication Technologies Conference (ICTC), Nanjing, China, 2023, pp. 243-247, DOI: 10.1109/ICTC57116.2023.10154831.

[6] C. Jiang, J. Shen, S. Chen, Y. Chen, D. Liu, and Y. Bo, "UWB NLOS/LOS Classification Using Deep Learning Method," in IEEE Communications Letters, vol. 24, no. 10, pp. 2226-2230, Oct. 2020, DOI: 10.1109/LCOMM.2020.2999904.

[7] A. H. Ismail, A. H. M. Hajazi, M. S. M. Azmi, M. S. M. Hashim, M. N. Ayob, and H. H. Basri, "Analysis of Wi-Fi Spatial-Temporal Data for Organic Fingerprinting-based Indoor Positioning System," 2022 2nd International Conference on Electronic and Electrical Engineering and Intelligent System (ICE3IS), Yogyakarta, Indonesia, 2022, pp. 130-134, DOI: 10.1109/ICE3IS56585.2022.10010207.

[8] Guo, G., Wang, H., Bell, D., Bi, Y., Greer, K. (2003). KNN Model-Based Approach in Classification. In: Meersman, R., Tari, Z., Schmidt, D.C. (eds) On the Move to Meaningful Internet Systems 2003: CoopIS, DOA, and ODBASE. OTM 2003. Lecture Notes in Computer Science, vol 2888. Springer, Berlin, Heidelberg. https://doi.org/10.1007/978-3-540-39964-3_62.

[9] W. Liu, "An Improved Indoor Location Algorithm for Wi-Fi Position Fingerprint Based on IABC-KMC," 2023 IEEE 3rd International Conference on Information Technology, Big Data and Artificial Intelligence (ICIBA), Chongqing, China, 2023, pp. 529-534, DOI: 10.1109/ICIBA56860.2023.10164873.

[10] M. Ezuma, F. Erden, C. Kumar Anjinappa, O. Ozdemir, and I. Guvenc, "Detection and Classification of UAVs Using RF Fingerprints in the Presence of Wi-Fi and Bluetooth Interference," in IEEE Open Journal of the Communications Society, vol. 1, pp. 60-76, 2020, DOI: 10.1109/OJCOMS.2019.2955889.

Journal of Physics: Conference Series **2813** (2024) 012004 doi:10.1088/1742-6596/2813/1/012004

# Line optimization of multi-beam sonar sounding technology in ocean surveying and mapping

**Mengjia Li[1*], Zien Zhou[2], Shaowei Lyu[3], Mengxuan Liang[4], Zixuan Gao[2] and Chengming Xia[2]**

[1]School of Mechanical and Electrical Engineering, Chengdu University of Technology, Chengdu, 610059, China
[2]Faculty of Science, Beijing University of Technology, Beijing, 100124, China
[3]School of Management, Guangdong University of Technology, Guangzhou, 510630, China
[4]Business School, Central South University, Changsha, 410083, China

[*] Corresponding author's e-mail: 2573414579@qq.com

**Abstract:** In this paper, the problem of line optimization existing in the practical application of multi-beam sonar-sounding technology is studied. Firstly, by analyzing the working principle of multi-beam sonar, the multi-beam coverage width model and the overlap rate model between adjacent strips are established considering the slope of seabed terrain. Then, a more general parameterized multi-beam coverage width formula with an adjustable azimuth angle of the line is derived under three-dimensional space coordinates. Then, the minimum total line length is defined as the objective function, and the constraint conditions of overlap rate and complete coverage are added to establish the mathematical model of line planning. Then, an iterative algorithm is designed to quickly determine the line spacing, and an improved particle swarm optimization algorithm, including a trigonometric function change learning factor, is used to search for the optimal line layout. Finally, the simulation results show that the optimized scheme can reduce the total line length by 2% compared with the unoptimized line layout, which verifies the effectiveness of the established model and algorithm. This study provides a theoretical basis for navigation planning and shipboard system optimization of multi-beam sonar survey. Subsequent work taking into account the influence of complex sea conditions, efforts were made to expand the model's applicability.

## 1. Introduction

With the rapid development of the marine industry, obtaining accurate seabed topography information to support marine resource development, shipping, national defense, and other needs has been mentioned at the forefront. Compared with the traditional single-beam sonar technology, multi-beam sonar technology simultaneously transmits multiple acoustic beams through a single probe and uses the phased array principle to rapidly detect and map the three-dimensional terrain of a large area of sea in a short period of time [1]. However, there are many problems in the engineering application, such as large redundancy and high data repetition rate, which not only reduces the work efficiency but also increases the ineffective navigation cost. Therefore, the study of line planning and optimization of multi-beam sonar has become the key to improving the accuracy and efficiency of ocean mapping[2].

## 2. Model building

*2.1 Cover model considering the terrain*

Since the seafloor topography is not perfectly horizontal, we introduce the slope angle $\alpha$, the angle between the left and right beams of the survey line and the seafloor $\gamma$. On this basis, the coverage model is shown in Figure 1:

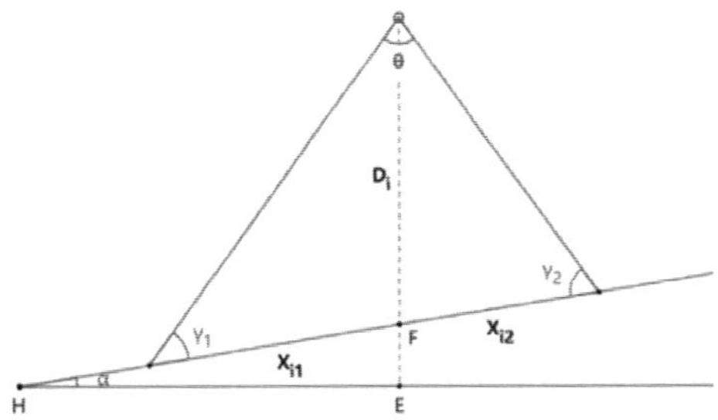


**Figure 1.** Multi-beam coverage model considering terrain.

For a single survey ship, the coverage width of the multi-beam survey is divided into $X_{i1}$ and $X_{i2}$ by the vertical line of the survey ship, and then the relationship between the required Angle and the submarine slope is established by using the method of calculating geometry and triangular relationship. Then, the sine theorem is used to establish the relationship between cover width, depth $D_i$, and seabed slope[3].

The length $X_{i1}$ and $X_{i2}$ of the left and right beams on the slope can be calculated by this equation:

$$X_{i1} = \frac{D_i \sin\frac{\theta}{2}}{\sin\gamma_1}, \quad X_{i2} = \frac{D_i \sin\frac{\theta}{2}}{\sin\gamma_2} \tag{1}$$

*2.2 Overlapping ratio model between adjacent strips in multi-beam sounding*

Analyzing the definition of the overlap rate in the plane shows that the overlap rate is essentially the length of the overlapping portion over the width covered by the beam. Therefore, the overlap rate when there is a slope is defined as the ratio of the overlap width between the current coverage and the last coverage to the current coverage width, where $C_i$ is the length of the overlapping part on the slope:

$$\eta_i = \frac{C_i \cos\alpha}{W_i} \tag{2}$$

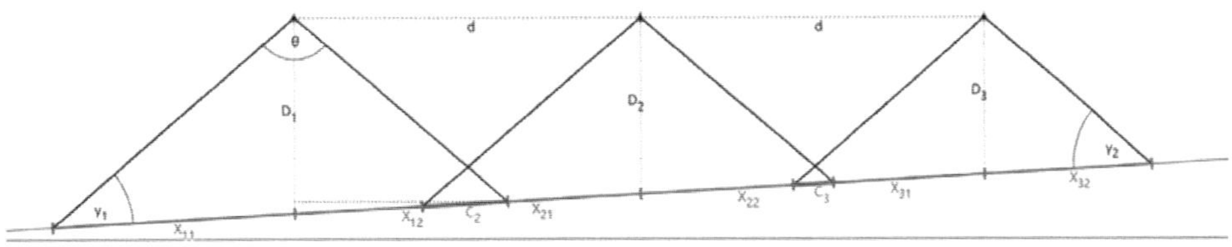


**Figure 2.** Overlay length diagram.

Figure 2 shows that $C_i = X_{i1} + X_{(i-1)2} - \frac{d}{\cos\alpha}$, and substituting $C_i$ and $W_i$ into the formula yields:

$$\eta_i = \frac{X_{i1} + X_{(i-1)2} - \frac{d}{\cos\alpha}}{X_{i1} + X_{i2}} = 1 - \left( \frac{\cos\alpha \cos\frac{\theta}{2}}{\cos(\frac{\theta}{2} - \alpha)} \right) \frac{d}{W_i} \tag{3}$$

### *2.3 Universal 3D coverage model*

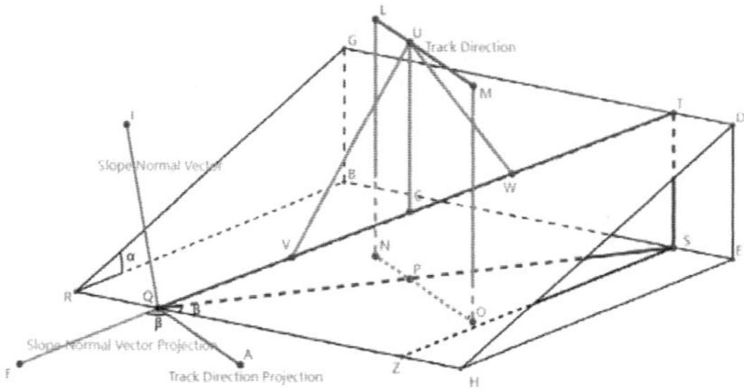


**Figure 3.** 3D model diagram.

As can be seen from Figure 3, $LM$ is the direction of the route, and its projection on the horizontal plane is $NO$. By translating it to $QA$, the projection from the direction of the $Q$ route can be obtained. $QI$ is the normal vector of slope $GRHD$, the normal aspect of submarine slope is projected as $QF$ on the horizontal plane, and the Angle between $QF$ and $QA$ is $\beta$. We make a plumb line in the direction of the course through a point on the course, and on the course, the angle of the seafloor slope corresponding to the transmitting detection wave is $\alpha_\beta = \angle TQS$. Since $AQ \perp QS$ $FQ \perp QH$, we have $\angle FQA = \angle SQH = \beta$. Then, we need $\alpha_\beta$ and $D_{R\beta}$ to solve the coverage width using the model[4].

It is then possible to calculate the angle $\alpha_\beta$ of the seafloor slope corresponding to an angle $\beta$ between the direction of the survey line and the projection of the normal direction of the seafloor slope on the horizontal plane:

$$\tan\alpha_\beta = \frac{|TS|}{|QS|} = \sin\beta\tan\alpha \tag{4}$$

### *2.4 Seawater depth calculation*

Through Figure 4, we calculate a $D_{R\beta}$:

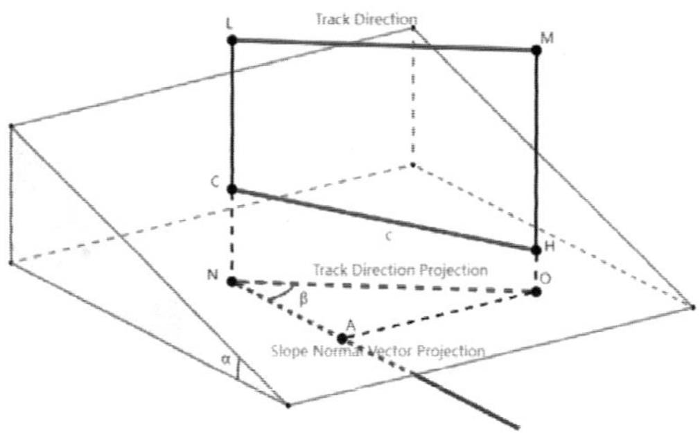


**Figure 4.** Schematic diagram of deriving $D_{R\beta}$.

The change of sea depth caused by the fall of the seabed slope $R\cos\beta$ represents the projection of the length of the fall of the survey ship along the slope's normal vector projection direction on the horizontal plane. That's the length of $NA$ in the picture. $R\cos\beta\tan\alpha$ is the height at which the depth of the sea drops, and this height can be calculated by using the tangent function in a right triangle, where angle $\alpha$ is the slope and $R\cos\beta$ is the adjacent side of the right triangle. We can get the general formula of $D_{R\beta}$:

$$D_{R\beta} = D_0 + R\cos\beta\tan\alpha \tag{5}$$

Finally, by substituting $D_{R\beta}$, the relationship between the coverage width and $R,\alpha,\beta,\theta$ is

$$W_{R\beta} = \frac{2\sin\theta(D_0 + R\cos\beta\tan\alpha)}{1+\cos\theta - \sin^2\beta\tan^2\alpha + \cos\theta\sin^2\beta\tan^2\alpha} \tag{6}$$

## 3. Algorithm design

### *3.1 Longitudinal wiring*

The longitudinal survey line layout problem can be transformed into a linear programming model,[5] assuming that there are n survey lines to be arranged, and the goal is to minimize the total length of survey lines when the detection range completely covers the given sea area. By traversing n in a certain range and finding the smallest n that gives the linear program a solution, the shortest length of the line can be found under the line layout scheme.

We need the following constraints to ensure that the detection width of the left beam on the first survey is greater than or equal to the distance from the first line, i.e., the distance from the last line to the sea area to be surveyed. The second condition is to ensure that the seafloor slope covered by the right beam of the last line intersects with the boundary of the sea area to ensure that the edge of the area is covered, and that the overlap is in the range of 10 to 20 %.

$$X_{11}\cos\alpha = \frac{D_1\sin\frac{\theta}{2}\cos\alpha}{\sin\gamma_1} \geq d_1 \tag{7}$$

$$\sum_{i=n}^{n} d_i + X_{n2}\cos\alpha \geq 4\times 1852 \tag{8}$$

$$\sum_{i=1}^{n} d_i \leq 4\times 1852 \tag{9}$$

$$0.1 \leq \eta_i \leq 0.2 \tag{10}$$

Using Python to solve linear programming, the optimal layout scheme of survey lines is obtained as follows: total number of survey lines: 34; total length of survey lines: 125,936 m (68 nm). Then, through simulation, Figure 5 is generated.

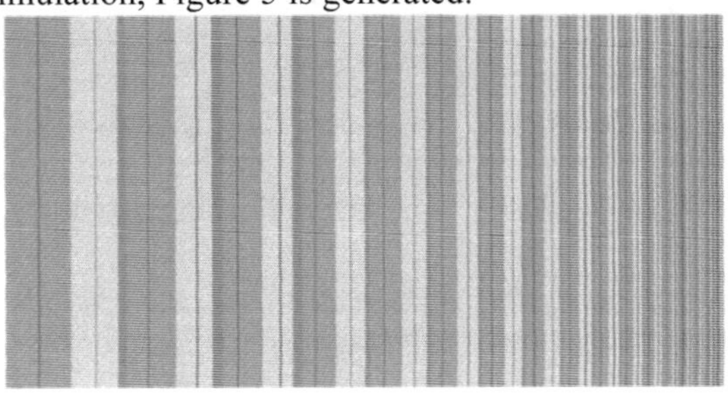

**Figure 5.** Longitudinal wiring.

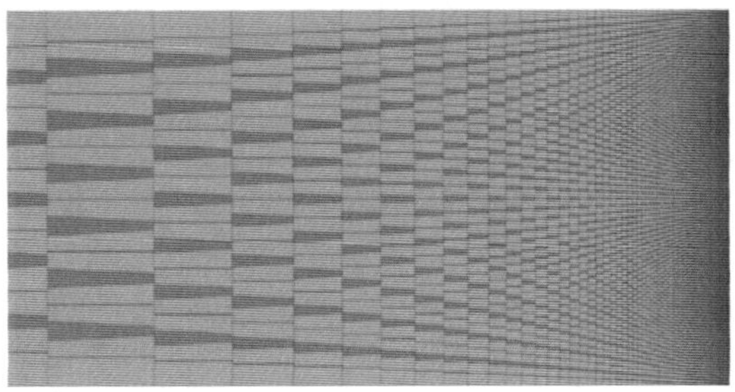

**Figure 6.** Horizontal wiring.

After analyzing the results of Figure 6, it is found that the overlap rate of one line and the previous one is 19%. The linear programming is considered to be optimized, and the transverse line is added. However, the number of measurement lines required for different horizontal coordinates of the horizontal wiring is different. If the measurement lines can be discontinuous, the optimal strategy for calculating the horizontal wiring can be calculated by sectional wiring, as shown in the following figure, where red is the measurement line and dark blue is the repeated measurement area. On the basis of this figure, the longitudinal rays are inserted from the west, and the dynamic programming optimization model is established to obtain the minimum line length.

## 4. Experimental results

*4.1 The solution of overlapping ratio model between adjacent strips in multi-beam sounding*

According to the above model, the seawater depth, covering width, and overlap rate of different measuring line distances from the center can be obtained. The results as shown in table 1.

**Table 1.** Calculation results.

| Distance from survey line to center/m | -800 | -600 | -400 | -200 | 0 | 200 | 400 | 600 | 800 |
|---|---|---|---|---|---|---|---|---|---|
| Sea depth /m | 90.95 | 85.71 | 80.47 | 75.24 | 70 | 64.76 | 59.53 | 54.29 | 49.05 |
| Covering width /m | 315.71 | 297.53 | 279.35 | 261.17 | 242.99 | 224.81 | 206.63 | 188.45 | 170.27 |
| The overlap rate with the previous line /% | —— | 35.7 | 31.52 | 26.74 | 21.26 | 14.89 | 7.41 | -1.53 | -12.36 |

If the opening Angle of the multi-beam converter is $120^{\circ}$, the slope is $1.5^{\circ}$, and the depth of the sea at the center point is 70 m, the known parameters are as follows: $\theta = 120^{\circ}$, $\alpha = 1.5^{\circ}$, $d = 200$, $D_5 = 70$. The seawater depth, covering width, and overlap rate with the previous line can be obtained using the model input parameters.

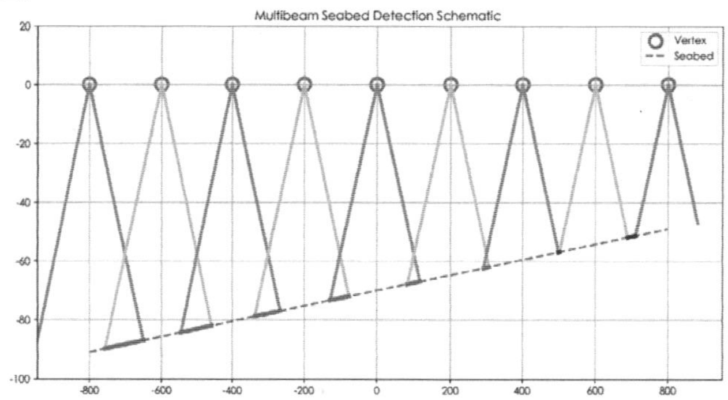


**Figure 7.** Results.

Figure 7 illustrates the correlation among beam transmitting points, beams, and detection ranges. Hollow black circles depict sea-level beam transmitting points, while submarine slopes are shown as black dashed lines. Each transmitting point emits two beams symmetrically, spanning 120 degrees to the left and right toward the sea floor. The line segment connecting the beams and the seabed slope intersection denotes their detection range. If adjacent detection ranges overlap, a thick red line signifies the overlap; otherwise, a thick blue line indicates the gap. This figure visually demonstrates the coincidence of detection ranges emitted from beam transmitting points.

*4.2 Solution of multi-beam sounding coverage width model in three-dimensional space*

If $\theta = 120^{\circ}$ θ, $\alpha = 1.5^{\circ}$, $D_0 = 120$. The coverage width can be obtained by substituting the parameters with the model. In the following table 2, according to the requirements of the topic, the coverage width can be obtained by bringing the above-known conditions into the above mathematical model.

**Table 2.** The calculation results of the model.

| Covering width /m | | Measure the distance of the ship from the center of the sea/ Sea mile | | | | | | | |
|---|---|---|---|---|---|---|---|---|---|
| | | 0 | 0.3 | 0.6 | 0.9 | 1.2 | 1.5 | 1.8 | 2.1 |
| Measurement line direction Angle /° | 0 | 415.55 | 465.93 | 516.31 | 566.69 | 617.08 | 667.46 | 717.84 | 768.22 |
| | 45 | 415.62 | 451.25 | 486.88 | 522.51 | 558.15 | 593.78 | 629.41 | 665.04 |
| | 90 | 415.69 | 415.69 | 415.69 | 415.69 | 415.69 | 415.69 | 415.69 | 415.69 |

Journal of Physics: Conference Series **2813** (2024) 012004 doi:10.1088/1742-6596/2813/1/012004

| | | | | | | | | |
|---|---|---|---|---|---|---|---|---|
| 135 | 415.62 | 379.99 | 344.36 | 308.72 | 273.1 | 237.46 | 201.83 | 166.2 |
| 180 | 415.55 | 365.17 | 314.79 | 264.41 | 214.02 | 163.64 | 113.26 | 62.88 |
| 225 | 415.62 | 379.99 | 344.36 | 308.72 | 273.1 | 237.46 | 201.83 | 166.2 |
| 270 | 415.69 | 415.69 | 415.69 | 415.69 | 415.69 | 415.69 | 415.69 | 415.69 |
| 315 | 415.62 | 451.25 | 486.88 | 522.51 | 558.15 | 593.78 | 629.41 | 665.04 |

*4.3 Solution of line model based on particle swarm optimization with improved learning factor*

The steps are as follows: First, we initialize particles with random positions (X) and velocities (V). Then, we calculate the fitness of each particle. Each particle compares its fitness with its own best position (Pi) and the global best position (Pg). The particle's velocity and position are updated using the PSO method formula. The algorithm optimizes the angle (β) and distance (d) between survey lines to meet sea area constraints and minimize overlap to 20%.

The improved particle swarm optimization algorithm adjusts β and d values based on each particle's fitness, aiming to minimize survey line length while covering the entire sea area with minimal overlap between adjacent strips. The structure diagram is shown in Figure 8.

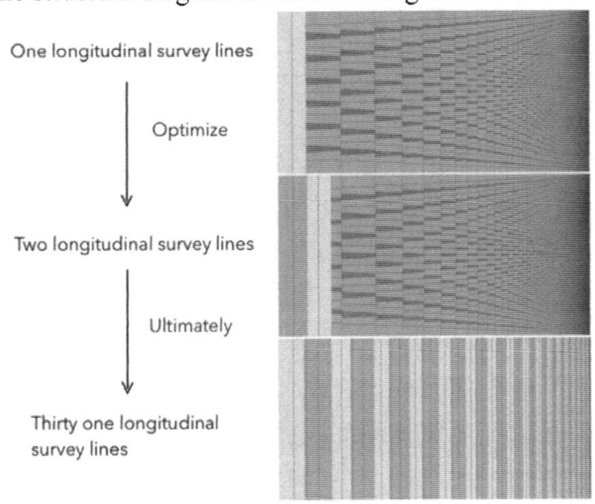


**Figure 8.** Iterative solution steps.

*4.4 Realistic undersea terrain*

The topography of the submarine topographic map can be obtained, and 11 rectangular areas can be divided accordingly. We drew three-dimensional maps of the 11 rectangular areas respectively to observe their topographic characteristics and divided them into two categories, as shown in the figure 9 and figure 10, one of which is an approximately horizontal plane, and the other is an approximately inclined plane.

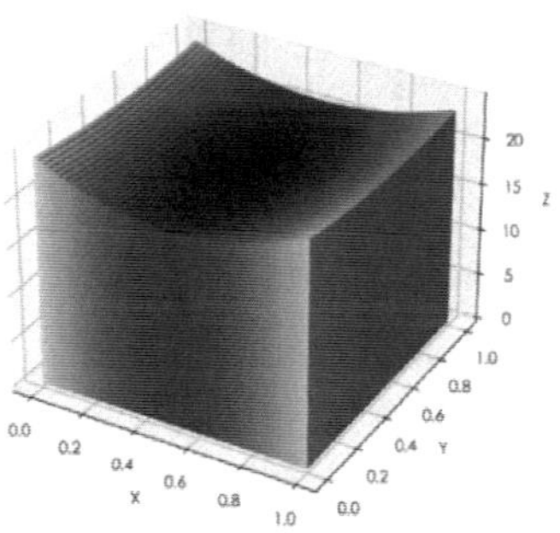


**Figure 9.** Area 1.

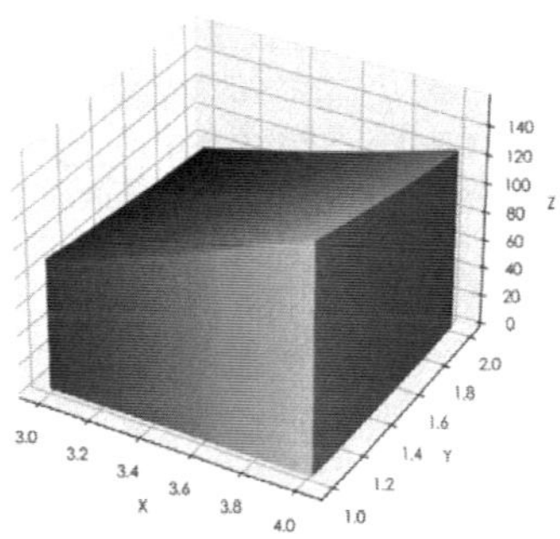


**Figure 10.** Area 2.

Using the previous linear programming model, the optimal measuring line scheme and the corresponding minimum values are obtained, as shown in Table 3.

**Table 3.** Solution results.

| | 1 | 2 | 3 | 4 | 5 | 6 | 7 | 8 | 9 | 10 | 11 | L/ Sea mile |
|---|---|---|---|---|---|---|---|---|---|---|---|---|
| 120 degrees | 21 | 28 | 43 | 17 | 17 | 10 | 7 | 10 | 6 | 9 | 5 | 245 |

The real seafloor line layout is shown in the Figure 11:

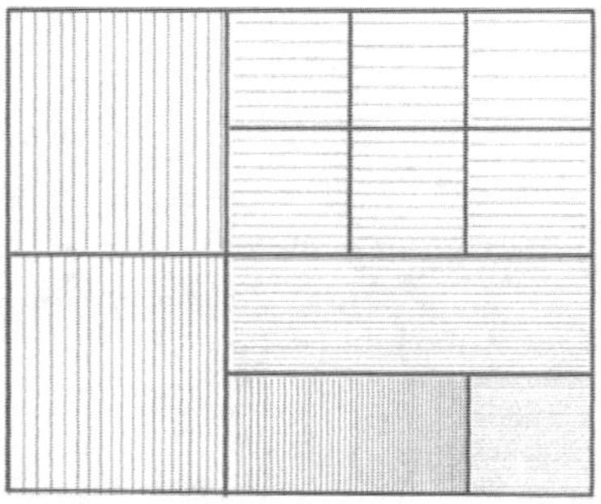

**Figure 11.** 5 × 4 sea area map.

## 5. Conclusion

Aiming at the redundancy of line layout in multi-beam sonar measurement, this paper achieved efficient line optimization by establishing a dynamic coverage width model, constructing an adjustable parametric 3D programming model, design iteration, and improved particle swarm optimization. Simulation verified the effect of the proposed method and achieved significant improvements of a 2% reduction in total length. The proposed model and algorithm provide practical guidance for the performance optimization of multi-beam sonar equipment and the solution to the problem of line optimization during use.

## References

[1] Wang Ning, Huang Jing, Rao Tiantian. Line optimization model in the case of multi-beam sounding system [J]. Journal of electroacoustic technology, 2023, 47 (6): 61-64. The DOI: 10.16311 / j.a udioe. 2023.06.018.

[2] Wang Junsen, Jin Shaohua, Bian Zhigang et al. Correction of residual of roll motion in multi-beam sounding using overlapping area of adjacent survey lines [J]. Journal of Marine Technology, 2019, 42(04):35-42.

[3] Jiao Anlong, Zou Yonggang. Research on technical design content of multi-beam survey in sea area [J]. Marine Surveying and Mapping, 2015, 35(03):43-45+49.

[4] Yu S, Yuqing C, Shengxiang Y, et al.Dynamic niching particle swarm optimization with an external archive-guided mechanism for multimodal multi-objective optimization [J]. Information Sciences, 2024, 653

[5] Rajashekar J K, Channakrishnaraju, Gowda C P, et al.SCEHO-IPSO: A Nature-Inspired Meta Heuristic Optimization for Task-Scheduling Policy in Cloud Computing [J]. Applied Sciences, 2023, 13(19).

Journal of Physics: Conference Series **2813** (2024) 012005 doi:10.1088/1742-6596/2813/1/012005

# Logical Mapping-based hierarchical safe reinforcement learning for autonomous driving

**Jihui Nie*** **and Yingda Li**

Software Engineering Institute, East China Normal University, Shanghai, 200062, China

*Corresponding author's e-mail: 51215902070@stu.ecnu.edu.cn

**Abstract:** Making decisions in complex and multiple scenarios presents great challenges for autonomous driving systems (ADS). In recent years, deep reinforcement learning algorithms (DRL) have made remarkable breakthroughs in decision-making. However, there remain many problems, such as sparse reward and slow convergence in traditional DRL when facing multiple sub-goals. In this paper, we propose a hierarchical deep deterministic policy gradient (DDPG) based on the BDIK model for autonomous driving, which enables ADS to have the ability to make decisions in human-like deliberation ways as well as deal with uncertainties in the environment. First, we propose the BDIK model based on the Beliefs-Desires-Intentions (BDI) model so that the agents are guided by domain knowledge when generating their sub-goals. Furthermore, in contrast to traditional BDI systems making plans by hand, a BDIK hierarchical DDPG (BDIK HDDPG) algorithm is employed to deduce the optimal actions automatically in an uncertain environment. The results show that our method outperforms the standard DDPG for both processing speed and effectiveness in multiple and complex scenarios.

## 1. Introduction

In the realm of intelligent transportation, autonomous driving decision-making is crucial, focusing on actions based on environmental perceptions [1]. Challenges include mimicking human decision-making, incorporating experience and regulations [2], and navigating uncertainties [3]. The Beliefs-Desires-Intentions model addresses planning and execution but struggles with complex scenarios [4, 5]. Deep Reinforcement Learning (DRL) [6] adapts to uncertainties but often lacks human knowledge integration. We introduce the BDIK HDDPG, merging BDIK reasoning for human-like deliberation with a hierarchical DDPG [7] algorithm for optimal action in uncertain conditions.

## 2. Method

Aiming at multi-task long-term autonomous driving, the HDDPG based on logic Mapping is proposed, and the logic Mapping is realized by the BDIK reasoning model. The overall framework is shown in Figure 1.

### *2.1 BDIK reasoning model*

In the BDI architecture [8], agents possess beliefs, desires, and intentions, where beliefs are sensory perceptions of the environment, desires represent agent objectives (e.g., Lane Following, Avoiding Collision), and intentions denote current goals. To streamline decision-making, deliberation involves choosing desires and determining means to achieve them; in our approach, this process involves reasoning from beliefs to desires, facilitated by driving experience rules generated from expert trajectory data via a random forest algorithm, followed by evaluating desires for security, legality, and effectiveness to establish the current intention. Meanwhile, means-end reasoning is left to reinforcement learning to achieve committed objectives.

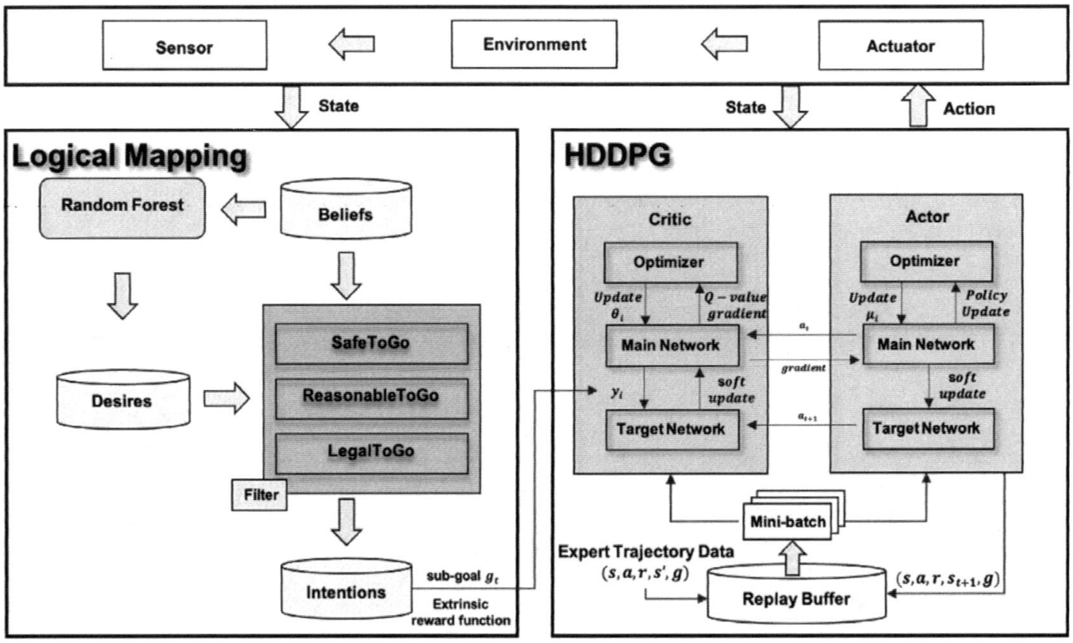


**Figure 1.** The overall framework.

### *2.2 Knowledge-based deliberation process*

Our research improves intelligent systems' decision-making by using the random forest [9] to derive rules from expert driving data. These rules help the system decide on its next actions. When faced with multiple options, we evaluate these against three essential criteria to ensure the chosen action is optimal:

- Safe To Go: Assessing if the action avoids potential hazards, using the Time to Collision (TTC) metric [10].
- Reasonable To Go: Determining whether the action moves the system closer to its destination.
- Legal To Go: Ensuring the action complies with traffic laws, like speed limits and obeying traffic signals.

### *2.3 BDIK reasoning cycle*

The perception module receives input from the sensor and updates beliefs, providing an agent with an understanding of the world. Then, the desires set are derived from the driving experience rule base constructed by the random forest algorithm. In the next section, we evaluate each of the desires set from three aspects: safety, legality, and rationality. Then, the best one is selected as the current intention. During this process, we assess whether the intention has been completed and whether a more urgent intention has emerged. After that, HDDPG receives the current intention as its sub-goal and returns an optimal action sequence to achieve the committed states of affairs. This process is repeated until the ego vehicle reaches its final destination.

### *2.4 BDIK HDDPG*

In order to achieve the inferred intention, we propose BDIK HDDPG. G is the set of all possible goals, and each goal $\boldsymbol{g} \in G$ corresponds to a specific reward function r: $S \times A \rightarrow R$. At the beginning of each episode, the BDIK reasoning model generates the sub-goal committed to do, and it is converted into a one-hot vector. The objective of the agent is to find an optimal goal-conditioned policy $\boldsymbol{\pi}^*$ which maximizes the action-value function:

$$Q_\pi(s, g, a) = \mathbf{E}_\pi([\textstyle\sum_{t=0} \gamma^t R(s_t, a_t, g_t) \,|s_t = s, g_t = g, a_t = a]). \quad (1)$$

The BDIK HDDPG comprises an actor network and a critic network. The actor network, denoted as $\boldsymbol{\pi}$: S × G → A, serves as the policy network and is updated using policy gradient (2).

$$\nabla_\varphi J(\varphi) = \frac{1}{N}\sum_i^N \nabla_a Q(s, a, g|\theta^Q)|_{s=s_i, a=\pi_{\varphi(s_i, g_i)}} \nabla_\varphi \pi_\varphi(s, g)|_{s_i, g_i} \quad (2)$$

where $Q(s, a, g|\theta^Q)$ is from the critic network, and N is the total number of mini-batch.
The critic network evaluates the Q-value function $Q(s, a, g|\theta)$ to criticize the actions made by the actor network based on the current state and goal. The critic network is updated based on:

$$L = \frac{1}{N}\sum_i^N (Q(s, a, g|\theta^Q - y^i))^2 \quad (3)$$

where $y^i = r_i + \gamma Q'(s', a_t, g_t|\theta^{Q'})$ is the Q-value estimated by the target network.

Similar to DDPG, we introduce a target critic network with parameter $\theta'$ and a target actor network with parameter $\varphi'$. The parameters of the target networks are updated in a fixed frequency as follows:

$$\begin{aligned} \theta' &= \tau\theta + (1-\tau)\theta' \\ \varphi' &= \tau\varphi + (1-\tau)\varphi' \end{aligned} \quad (4)$$

We store the agent's experiences at each time step and randomly sample a mini-batch of transitions from the replay buffer. In order to make the agent better learn from the expert experience, we use expert trajectory data to create quintuple transitions $(s, a, r, s', g)$, and store them in the replay buffer. Therefore, the agent can obtain successful experience in the learning stage, encouraging the agent to actively perform certain behaviors.

We firstly initialize main actor network $\mu(s, g|\varphi)$ and main critic network $Q(s, a, g|\theta)$ as well as target network $\mu'(s, g|\varphi')$ and $Q'(s, a, g|\theta')$ with weights $\theta' \leftarrow \theta$ and $\varphi' \leftarrow \varphi$. We then construct an empty replay buffer B with max memory length $l_b$. In the training stage, we first get the current goal from the BDIK reasoning model. If the maximum number of attempts H is not reached, we judge whether the current goal has been achieved. If the goal has been achieved, then we re-select the goal; otherwise, we hand it over to HDDPG as the committed state of affairs. The main actor network interacts with the driving environment. More specifically, it receives the current state $s_t$ and goal $g_t$, if it is not a terminate state, it generates an action $a_t = \mu(s_t|\theta_t, g_t)$. Due to the specificity of the different autonomous driving tasks, instead of evaluating all the final outputs of the actor with one reward function, we designed specific reward functions for different driving tasks. According to the driving task the BDIK model inferred, the corresponding reward function is activated, the current reward $r_t$ is returned, and the next state $s_{t+1}$ is given. Then, the transition tuple $(s_t, a_t, r_t, s_{t+1})$ s stored in the replay buffer. At the same time, we store the "expert experience" obtained in advance from the expert trajectory data into the replay buffer, which helps agents better learn from expert experience.

## 3. Experiment

### 3.1. Experiment settings

We chose the CARLA simulator [11] to perform all our experiments, where dynamic obstacles may appear in random positions. As shown in Figure 2, the two sub-goals are adaptive cruise control (ACC) and cross intersection (CI).

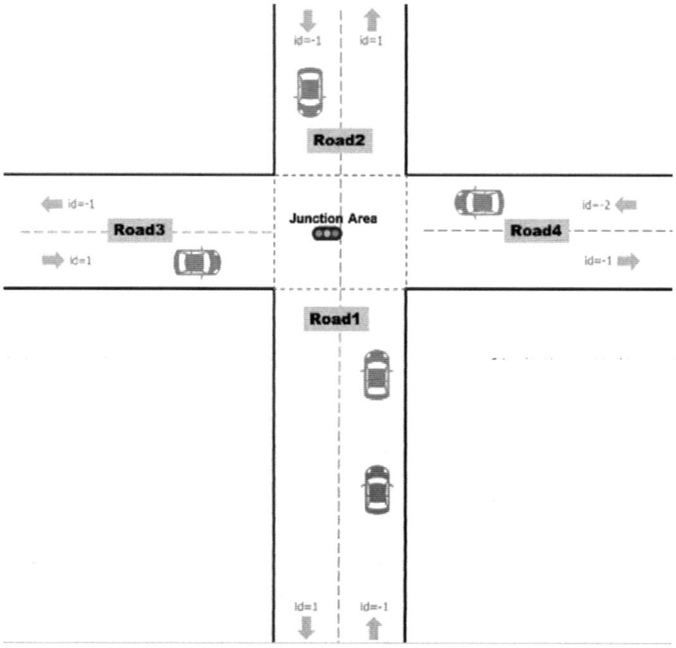


**Figure 2.** Self-driving car approaching intersection with traffic light.

The state space is a 9-dimensional tuple:

$$S = \left[v_e, v_{des}, \varphi, d_{lane}, d_f, d_{des}, d_{safe}^{v}, s_{light}, d_{sl}\right] \tag{5}$$

Equation (5) is the state space, where $v_e$ and $v_{des}$ represent the actual speed and desired speed of the ego car, respectively. $\varphi$ and $d_{lane}$ are respectively the angle and distance between the ego car and the road axis. $d_f$, $d_{des}$, and $d_{safe}^{v}$ denote the actual distance, desired distance, and safe distance from the ego car to the nearest front vehicle. In addition, we introduce a traffic light signal $s_{light}$ and actual distance $d_{sl}$ between the ego car and the stop line.

The BDIK model outputs the selected sub-goal: ACC or CI. The action space is formulated as a 2-dimension vector for throttle $\in [-3, 3]$ and steering $\in [-0.3, 0.3]$ respectively.

Assume that for one step, the selected sub-goal is denoted as $g_t$. And each sub-goal corresponds to an extrinsic reward. For ACC, the reward function is given by Equation (6), where $r_1^{ACC}$ encourages driving in the center of the road, and $r_2^{ACC}$ hopes to travel at the desired speed. $r_3^{ACC}$ requires the ego car to maintain a safe distance from the front vehicle. If a collision occurs, then we terminate and give a penalty of -100.

$$\left\{\begin{matrix} r_1^{ACC} = -k_1 sin\varphi - k_2 abs(d_{lane}) \\ r_2^{ACC} = -k_3 abs(d - d_{des}) \\ r_3^{ACC} = \left\{\begin{matrix} -k_2 abs(d - d_{safe}), if\ not\ collide \\ -100, if\ collide \end{matrix}\right\} \end{matrix}\right\} \tag{6}$$

For CI, the reward function is given by Equation (7). In this case, the ego car is required not to deviate from the lane, and it is penalized when traffic rules are violated.

$$\left\{\begin{matrix} r_1^{CI} = r_1^{c} = -k_1 sin\varphi - k_2 abs(d_{lane}) \\ r_2^{CI} = \left\{\begin{matrix} -100, if\ s_{light} = green\ and\ v = 0 \\ -100, if\ s_{light} = red\ and\ \dfrac{v^2}{2d_{st}} > a_{max} \\ -100, if\ collide \end{matrix}\right\} \end{matrix}\right\} \tag{7}$$

In this multi-task scenario, we first generate an expert trajectories data set from the CARLA simulator and then use the random forest algorithm package in WEKA version 3.7.11 to build the driving experience rule base automatically. With the help of the BDIK reasoning cycle, the current intention is generated. The sub-goal is converted into a one-hot vector and fed into H-DDPG as its sub-goal.

Because our input is low-dimensional temporal data, we use feedforward neural networks in our model. The structure of the actor network is composed of three liner layers and three fully connected layers, while the critic network contains three liner layers and two fully connected layers. In the actor network and critic network, all hidden layers contain 256 hidden units, and the layer is normalized. Except for the last layer of the actor network, which employed tanh as an activation function, all hidden layers use a modified linear unit (ReLU) activation function to prevent gradient explosion and disappearance. The output layer contains two units corresponding to acceleration and angular velocity, respectively. Table 1 describes the hyperparameter settings.

**Table 1.** Hyperparameters and values for the DDPG model.

| Hyperparameters | Values | Description |
|---|---|---|
| LRA | 0.0001 | Learning Rate of Actor Network |
| LRC | 0.001 | Learning Rate of Critic Network |
| URT | 0.001 | Update Rate of Target Network |
| $\gamma$ | 0.99 | Discount Factor |
| BGD_SIZE | 50 | Batch Gradient Descent Size |
| BUFFER_SIZE | 100000 | Replay Buffer Size |

*3.2. Experimental results and analysis*

In order to demonstrate the efficacy of our method, we compare the BDIK-HDDPG with the original DDPG algorithm. Both of them are trained for 600 episodes, and the returns are depicted in Figure 3. Based on the definition of the reward function, a higher return indicates better control action choices. It can be found that the training stability and learning speed of our method are better than the original DDPG. This shows the good performance of hierarchical DDPG under the guidance of the BDIK reasoning model.

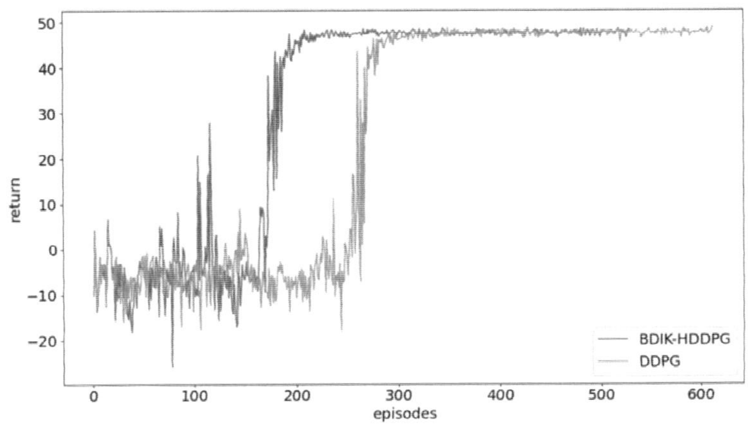


**Figure 3.** The comparison of return.

In addition, qualitative results comparing the trajectories of four different controllers are presented. Figure 4 shows the trajectories conducted by the expert, our method, the original DDPG, and the traditional BDI mechanism. The traditional BDI model selects hand-encoded plans corresponding to intentions, and the expert trajectories are generated by humans. It can be observed that the trajectory based on BDIK-HDDPG better imitates the human deliberation process.

Journal of Physics: Conference Series **2813** (2024) 012005 doi:10.1088/1742-6596/2813/1/012005

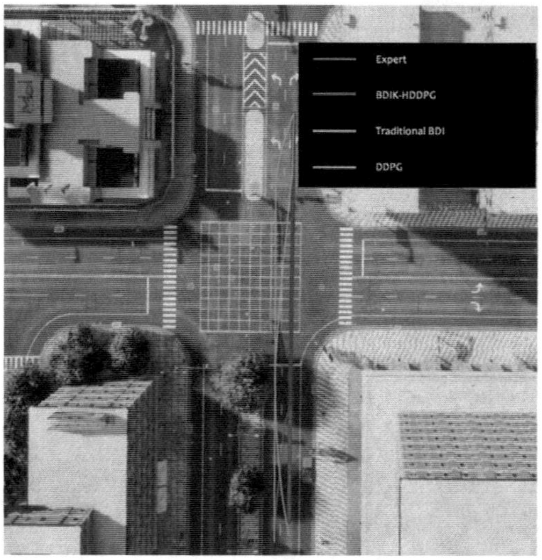


**Figure 4.** The comparison of trajectories.

In order to evaluate our approach's effectiveness in an unfamiliar dynamic environment, we compared different approaches across various tasks. In the simple scenario, both our method and the original DDPG showed a high success rate, while traditional BDI failed due to its inability to handle uncertainty. However, in the multiple-task scenario, the original DDPG resulted in more collisions and traffic violations. As shown in Table 2, our approach achieved obstacle avoidance with a high success rate in unknown environments and demonstrated good compliance with traffic rules.

**Table 2.** Comparison of models in deployment phases.

| **Task** | Methods | Collision | Break Traffic rules | Success |
|---|---|---|---|---|
| | DDPG | 4% | 3% | 93% |
| **ACC** | Tradition BDI | 20% | 5% | 75% |
| | BDIK-HDDPG | 3% | 2% | 95% |
| | DDPG | 5% | 6% | 89% |
| **CI** | Tradition BDI | 25% | 3% | 72% |
| | BDIK-HDDPG | 2% | 4% | 94% |
| | DDPG | 10% | 13% | 77% |
| **ACC + CI** | Tradition BDI | 45% | 6% | 49% |
| | BDIK-HDDPG | 5% | 3% | 92% |

## 4. Conclusion

In this paper, we propose a hierarchical DDPG framework based on the BDIK reasoning model for autonomous driving. The BDIK model generates sub-goals at the logical level, while the hierarchical DDPG algorithm automatically deduces optimal actions in uncertain environments. Experimental results demonstrate that our method enhances the average reward, accelerates the convergence rate of incentives, and establishes a model for adhering to traffic regulations. These outcomes contribute towards achieving our objective of emulating human thought processes.

## References

[1] Woodward, B., Kliestik, T.: Intelligent transportation applications, autonomous vehicle perception sensor data, and decision-making self-driving car control algorithms in smart sustainable urban mobility systems. Contemporary Readings in Law and Social Justice 13(2), 51–64 (2021). https://doi.org/10.22381/CRLSJ13220214.

[2] Hang, P., Lv, C., Xing, Y., Huang, C., Hu, Z.: Human-like decision making for autonomous driving: A noncooperative game theoretic approach. IEEE Transactions on Intelligent Transportation Systems 22(4), 2076–2087 (2020). https://doi.org/10.1109/TITS.2020.3036984.

[3] Shafaei, S., Kugele, S., Osman, M.H., Knoll, A.: Uncertainty in machine learning: A safety perspective on autonomous driving. In: International Conference on Computer Safety, Reliability, and Security. pp. 458–464. Springer (2018). https://doi.org/10.1007/978-3-319-99229-7_39.

[4] Alkire, S., Black, R.: A practical reasoning theory of development ethics: furthering the capabilities approach. Journal of International development 9(2), 263–279 (1997). https://doi.org/10.1002/(SICI)1099-1328(199703)9:2<263::AID-JID439>3.0.CO;2-D.

[5] Bratman, M.: Intention, plans, and practical reason (1987). https://doi.org/10.2307/2215590.

[6] Kiran, B.R., Sobh, I., Talpaert, V., Mannion, P., Al Sallab, A.A., Yogamani, S., P′erez, P.: Deep reinforcement learning for autonomous driving: A survey. IEEE Transactions on Intelligent Transportation Systems (2021). https://doi.org/10.1109/TITS.2021.3054625.

[7] Wang, S., Jia, D., Weng, X.: Deep reinforcement learning for autonomous driving. arXiv preprint arXiv:1811.11329 (2018). https://doi.org/10.48550/arXiv.1811.11329.

[8] Bratman, M.E., Israel, D.J., Pollack, M.E.: Plans and resource-bounded practical reasoning. Computational intelligence 4(3), 349–355 (1988). https://doi.org/10.1111/j.1467-8640.1988.tb00284.x.

[9] Frank, E., Hall, M., Holmes, G., Kirkby, R., Pfahringer, B., Witten, I.H., Trigg, L.: Weka-a machine learning workbench for data mining. In: Data mining and knowledge discovery handbook, pp. 1269–1277. Springer (2009). https://doi.org/10.1007/978-0-387-09823-4_66.

[10] Minderhoud, M.M., Bovy, P.H.: Extended time-to-collision measures for road traffic safety assessment. Accident Analysis & Prevention 33(1), 89–97 (2001). https://doi.org/10.1016/S0001-4575(00)00019-1.

[11] Dosovitskiy, A., Ros, G., Codevilla, F., Lopez, A., Koltun, V.: Carla: An open urban driving simulator. In: Conference on robot learning. pp. 1–16. PMLR (2017). https://doi.org/10.48550/arXiv.1711.03938.

Journal of Physics: Conference Series **2813** (2024) 012006
doi:10.1088/1742-6596/2813/1/012006

# Causal V: A five-phase causal mining methodology for real-world accident data

**Jiangnan Zhao***

Software Engineering Institute, East China Normal University, Shanghai, China

*Corresponding author's e-mail: Jnz956@foxmail.com

**Abstract.** Amid rapid advancements in autonomous vehicle technology, traffic accidents due to Autonomous Driving Systems (ADS) flaws persist. The critical task of attributing causality to accidents and forecasting potential incidents based on situational data stands at the forefront of enhancing vehicular safety. Addressing these challenges, this study introduces Casual V, a systematic five-phase V-shape approach to constructing a causal Bayesian network (CBN) from massive real-work crash data. To evaluate the proposed approach, a CBN was constructed from more than 300,000 Maryland accident records. Experiments demonstrate that our method is capable of effectively constructing a CBN that provides deep insights into the causal relationships within vehicular accident data.

## 1. Introduction

In recent years, with continuous breakthroughs and innovations in technology, the research and application of Autonomous Vehicles (AVs) have made significant progress. However, it cannot be overlooked that traffic accidents caused by defects in the Autonomous Driving Systems (ADS) still occur from time to time [1, 10]. It is of paramount importance to conduct causality analysis on traffic accidents based on vehicular accident data and to predict accidents according to the current scenario.

In response to the challenges, this paper proposes Causal V: a standardized five-phase V-shape construction process for a causal Bayesian network (CBN) [2] to support the extraction of causal relationships from real-world accident data, conduct retrospective causality analysis of the causes of accidents, and facilitate the prediction of traffic accidents based on the current scenario.

## 2. Method

As depicted in Figure 1, the 'Causal V' comprises five stages for constructing a Causal Bayesian Network. These stages are requirement collection, concept definition, development, selection & validation, and application. The following sections will introduce each of these stages.

### *2.1. Requirement collection phase*

During the requirement collection phase, the emphasis is placed on the collection of data pertinent to the objectives of the study, alongside a comprehensive assembly of application-specific requirements. For instance, this entails the meticulous analysis of historical traffic accident datasets to facilitate retrospective causality assessment of vehicular collisions.

In this context, the present research undertook a comparative evaluation of multiple accident datasets, ultimately selecting the Maryland crash data [3] as the most suitable source for this endeavor. This dataset serves as the foundation for conducting in-depth causality analysis of traffic accidents and underpins the development of predictive models for future incident forecasting.

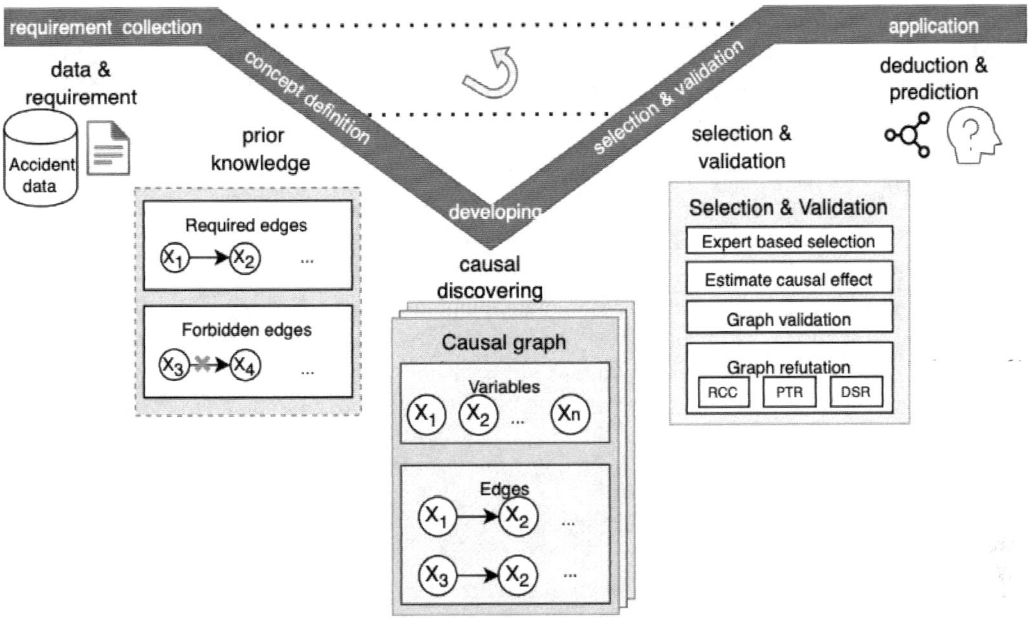


**Figure 1.** Overview of Causal V.

*2.2. Concept definition phase*

In the concept definition phase of the study, the incorporation of a priori knowledge is crucial for optimizing the exploration process of causal relationships. This phase necessitates the integration of domain-specific expertise, such as understanding the influence of temporal variations on light intensity. The infusion of such expert knowledge enables the algorithm to efficiently filter out potential causal structures, focusing on the most plausible causal connections. Moreover, by explicitly specifying forbidden and permissible edges within the model, the search space is significantly narrowed. This tailored approach not only enhances search efficiency but also augments the accuracy with which the algorithm constructs causal frameworks. Employing this strategy ensures that the algorithm can more swiftly and accurately unravel the authentic causal relationships embedded within the data.

*2.3. Developing phase*

In the development phase, a critical step is to pre-process the data amassed during the requirement collection stage, followed by the construction of preliminary causal models utilizing various causal discovery algorithms. The application of algorithms such as Greedy Fast Causal Inference (GFCI) [4] and Fast Causal Inference (FCI) [4] is pivotal in mining causal relationships within the data. Employing statistical measures like the G Square test, mutual information, and chi-square tests [5] is essential for detecting dependencies between variables. These statistical tools are integral for the meticulous refinement of the causal model, ensuring that the resultant structure is a robust and accurate representation of the underlying causal relationships.

Despite its ample volume of Maryland crash data, the dataset exhibited dispersed records with notable instances of missing data. Moreover, not all fields within the crash records were applicable for the construction of CBN. In pursuit of identifying factors genuinely influencing the severity of accidents, this study aggregated scattered crash data entries per incident, eliminating redundant or potentially confounding fields from the original dataset, such as car manufacturers, geographical coordinates of accidents, and county names. Following the data cleaning and filtering process, the constructed database encompassed **300,000 records**, with **33 dimensions** related to traffic accidents and a total of **212 attribute values**. This streamlined dataset lays the groundwork for a robust analysis, aiming to elucidate the underlying factors affecting accident severity.

Journal of Physics: Conference Series **2813** (2024) 012006 doi:10.1088/1742-6596/2813/1/012006

### *2.4. Selection & validation phase*

During the selection & validation phase, the preliminary causal models constructed in the preceding stage undergo a rigorous process of selection and validation. Initially, edges are filtered based on expert knowledge to refine the model's structure. Subsequently, the causal effects between variables are computed to further delineate the causal relationships. The validation of the causal graph is then performed using Structural Hamming Distance (SHD) [5] and Structural Intervention Distance (SID) [6], which serve as metrics to evaluate the accuracy and reliability of the inferred causal relationships. Finally, the causal graph is subjected to rebuttal through three distinct methodologies: Random Common Cause (RCC) [5], Placebo Treatment Refuter (PTR) [7], and Data Subset Refuter (DSR) [8].

Let $G(V, E)$ denote a causal model, where $V$ represents the set of all nodes, $V = \{V_1, V_2, \dots, V_n\}$, and $E$ constitute the collection of directed edges. The causal effect of node $V_i$ on node $V_j$ can be calculated with the use of do-calculus [9], which formalizes interventions within the causal model. The causal effect of intervening on $V_i$ to set its value to $v_i$ and observing the outcome on $V_j$ can be expressed as

$$Pr\left(V_j \mid do(V_i = v_i)\right), \tag{1}$$

where the $do(\cdot)$ operator signifies a controlled intervention that sets $V_i$ to a specific value, overriding its natural causes.

Furthermore, the Average Causal Effect (ACE), defined as

$$E\left[V_j \mid do(V_i = v_i)\right] - E\left[V_j \mid do(V_i =. v_i')\right], \tag{2}$$

where $v_i'$ is a baseline value of $V_i$, quantifies the expected change in $V_j$ as a result of changing $V_i$ from $v_i'$ to $v_i$.

SHD measures the difference between the estimated and the true causal graph by calculating the minimum number of edge operations (addition, deletion, or reversal) needed to transform the estimated graph to match the true graph. For two Directed Acyclic Graphs (DAGs) $G_1$ and $G_2$, SHD is defined as:

$$SHD(G_1, G_2) = COUNT(additions) + COUNT(deletions) + COUNT(reversals) \tag{3}$$

SID is designed to evaluate how well an estimated causal graph predicts the effects of interventions. SID calculates the number of variable pairs where the estimated graph and the true graph predict different causal effects. SID is defined as:

$$\text{SID}(G_1, G_2) = \sum_{(X,Y)\in V\times V} \mathbb{I}\{\text{if the effect of do}(X) \text{ on } Y \text{ differs in } G_1 \text{ and } G_2\} \tag{4}$$

The RCC method adds random covariates to the dataset and reassesses causal estimates to test stability; minimal change supports the initial hypothesis. PTR assigns a covariate as a placebo treatment, expecting a zero effect if the hypothesis holds, challenging the model's discernment of causal relationships. The DSR checks for consistency in causal estimates across data subsets, akin to cross-validation; stability suggests robust causality. These methods enhance confidence in the causal model without proving its correctness, by demonstrating resilience and reliability of causal estimates under various conditions.

### *2.5. Application phase*

In the application phase, the validated causal model from the previous stage is utilized for Bayesian parameter learning and structural learning to derive a Causal Bayesian Network (CBN). Once the CBN is successfully constructed, it facilitates a range of analysis and prediction tasks, including but not limited to the following aspects:

**Causal Inference**: One of the primary uses of a CBN is to perform causal inference, i.e., inferring the causal relationships between variables through observational data. This involves identifying which factors cause specific outcomes and how changes to these factors would affect the outcomes. This is particularly important for policy-making and intervention measures, as it provides a scientific basis for predicting the effects of different decision-making scenarios.

- **Causal Inference**: CBN can be used to infer the causal relationships between variables through observational data. This involves identifying which factors cause specific outcomes and how changes to these factors would affect the outcomes.
- **Decision Analysis**: CBNs can be employed to assess potential outcomes under different decision options, thereby aiding decision-makers in selecting the optimal course of action. By simulating the potential impacts of various strategies, CBNs offer clear guidance for complex decisions, especially in resource allocation, policy formulation, and risk management.
- **Predictive Modeling**: CBNs leverage existing causal relationships and conditional probabilities for prediction. This capability makes CBNs able to forecast future events or conditions, thus supporting the formulation of planning and preventive measures.
- **Intervention Analysis**: By simulating interventions in certain nodes of the causal network, CBNs can analyze the potential effects of specific intervention measures.

These applications underscore the versatility and practicality of CBNs in a wide range of analytical and predictive contexts, highlighting their importance in scientific research and decision-making processes.

## 3. Case study

In the case study section, this paper constructs a Causal Bayesian Network (CBN) based on Maryland crash data. Following the initial data pre-processing and the construction of a preliminary causal model, Table 1 presents the results of refutations conducted on the refined causal model. The result bolsters our confidence in utilizing this causal model.

**Table 1.** Refutation results.

| Method | Results | Confidence |
|---|---|---|
| RCC | Estimated effect: 1853.532<br>New effect: 1801.223 | High |
| PTR | Estimated effect: 1632.98<br>New effect: -211.347<br>p-value: 0.421 | High |
| DSR | Estimated effect: 1655.654<br>New effect: 1632.211<br>p-value: 0.231 | High |

After performing Bayesian structure and parameter learning on the causal model, the resulting CBN is illustrated in Figure 2. Arrows indicate causality directions, showing how variables like "Weather" directly affect "Road Conditions". The numbers in the model represent the probabilities of interactions between attributes, quantifying their causal relationships.

Utilizing this CBN, we can infer the cause of an accident. For example, we examine a rear-end collision occurring on a foggy morning, and we configure the accident scenario parameters within the Causal Bayesian Network (CBN) for Weather, Vehicle Movement, Lighting Conditions, and Accident Type as "foggy," "moving at a constant speed," "dawn," and "same direction rear end," respectively. This setup allows the CBN to infer the causes of traffic accidents under these conditions. The inference results indicate that foggy weather directly increases the likelihood of rear-end collisions, dawn's low light further aggravates visibility issues, and vehicles moving at high speeds may not decelerate in time due to limited reaction opportunities.

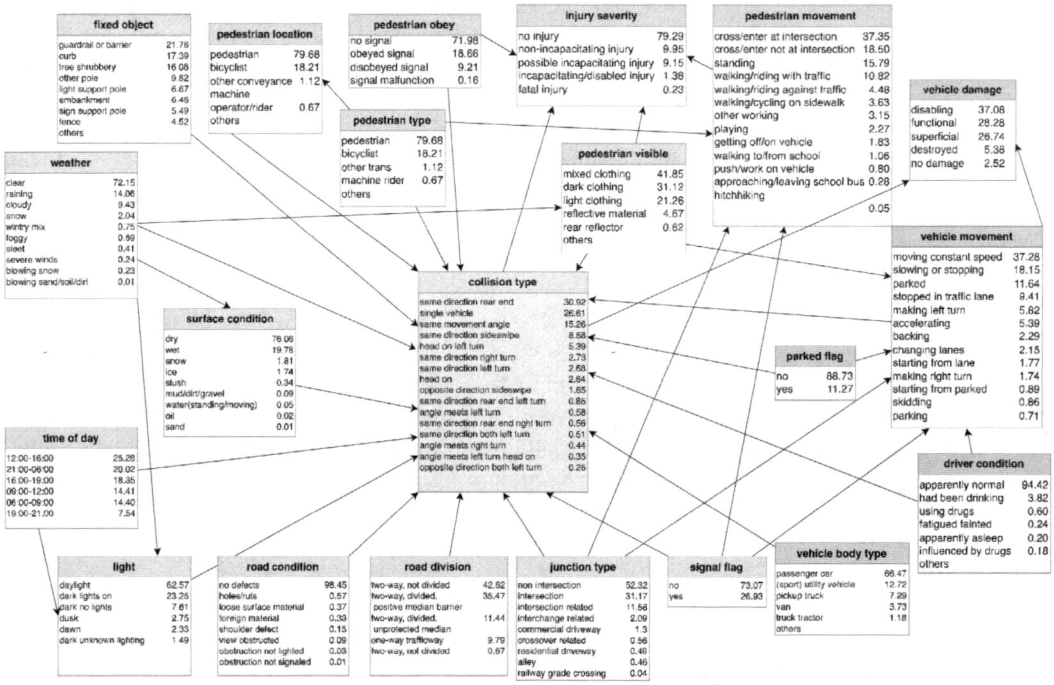


**Figure 2.** CBN constructed from Maryland accident data.

Leveraging the CBN for counterfactual reasoning, we can alter the weather condition to "clear" and re-run the model to deduce the potential occurrence and type of accidents. This hypothetical adjustment allows us to explore how changes in specific conditions could influence accident dynamics. The results of this counterfactual analysis can provide insights into how different scenarios might affect the probability of accidents, enabling more informed decisions regarding traffic safety measures and interventions.

## 4. Conclusion

This paper introduces "Causal V," a standardized five-phase V-shaped construction process for Causal Bayesian Networks (CBNs), designed to support the causal mining of real-world accident data. To evaluate the proposed approach, a CBN was constructed utilizing Maryland crash data. This model was applied to a specific accident scenario, performing both causality tracing and counterfactual reasoning. The analysis demonstrated the model's ability to identify underlying causal factors contributing to accidents and predict potential incidents under varying conditions. The results underscore the utility of "Causal V" in enhancing our understanding of accident causation and in developing more effective safety measures. Through this approach, we contribute to the field of traffic safety research by providing a robust tool for analyzing and predicting traffic accidents, thereby offering valuable insights for policy-making and preventive strategies.

## References

[1] Boudette, N.E., Metz, C. & Ewing, J., 2022. Tesla autopilot and other driver-assist systems linked to hundreds of crashes. https://www.nytimes.com/2022/06/15/business/self-driving-car-nhtsa-crash-data.html.

[2] Pearl, J., & Mackenzie, D. 2018 The Book of Why. Basic Books, New York.

[3] Maryland Benchmark Reports – Pages — mva.maryland.gov, 2024. [Accessed 11-01-2024].

[4] Nogueira, A.R., Pugnana, A., Ruggieri, S., Pedreschi, D. & Gama, J. 2022 Methods and Tools for Causal Discovery and Causal Inference. WIREs Data Mining and Knowledge Discovery, 12(2): e1449.

[5] Blöbaum, P., Götz, P., Budhathoki, K., Mastakouri, A.A. & Janzing, D. 2022. Do Why-GCM: An extension of Do Why for causal inference in graphical causal models. arXiv preprint arXiv:2206.06821.

[6] Peters, J. & Bühlmann, P. 2015. Structural intervention distance for evaluating causal graphs. Neural Computation, 27(3): 771-799.

[7] Naser, M.Z. & Çiftçioğlu, A.Ö. 2023. Causal discovery and inference for evaluating fire resistance of structural members through causal learning and domain knowledge. Structural Concrete. Wiley Online Library.

[8] Hason Rudd, D., Huo, H., and Xu, G., 2022. Improved Churn Causal Analysis Through Restrained High-Dimensional Feature Space Effects in Financial Institutions. Human-Centric Intelligent Systems, 2(3-4), pp.70-80.

[9] Pearl, J., 2012. The do-calculus revisited. arXiv preprint arXiv:1210.4852.

[10] Kim, S., Liu, M., Rhee, J.J., Jeon, Y., Kwon, Y., and Kim, C.H., 2022, November. Drivefuzz: Discovering autonomous driving bugs through driving quality-guided fuzzing. In Proceedings of the 2022 ACM SIGSAC Conference on Computer and Communications Security (pp. 1753-1767).

# Image segmentation based on improved hybrid krill herd algorithm

**Peng Zhang[1] and Ye Tian[1,2*]**

[1] College of Computer Science and Technology Changchun University of Science and Technology, Changchun, Jilin, 130022, China
[2] Zhongshan Research Institute of Changchun University of Science and Technology, Zhongshan, Guangdong, 528437, China

[*] Corresponding author's e-mail address: 13345105590@163.com

**Abstract.** Image segmentation is an important problem in the field of image processing. Traditional multi-threshold image segmentation faces the problems of inefficiency and lack of real-time performance. In order to improve these problems, a Kapur multi-threshold segmentation algorithm based on the improved hybrid krill herd algorithm is proposed, which adopts a novel population iterative approach limited to the top krill to effectively alleviate the phenomenon of excessive aggregation of krill. Building on this foundation, the differential evolutionary algorithm is combined with the krill herd algorithm, which provides a sufficient search range to strengthen the ability of the algorithm to jump out of the local optimum. The experimental results of multi-threshold segmentation on three classical test images show that the proposed improved algorithm can achieve higher segmentation accuracy and quality, and the convergence rate is mostly above 90%, which verifies the reliability and effectiveness of the algorithm in multi-threshold image segmentation problems.

## 1. Introduction

Image segmentation is one of the basic steps in the image processing process[1], which is able to extract multiple homogeneous regions based on the differences between pixels. The segmented image can be better applied in subsequent image processing work[2-3]. Among the current common image segmentation algorithms, threshold segmentation has the advantage of simplicity and efficiency. In practical problems, the target in the image is often not single, so the multi-threshold segmentation method has a broader range of applications. Kapur is a threshold segmentation method based on information entropy to maximize the entropy value of each image region, and it can provide better segmentation performance, which has been recognized by researchers[4]. The traditional threshold selection method (exhaustive method) can cope with the single-threshold segmentation problem, but the increase in the number of thresholds will lead to its inability to find the optimal segmentation threshold quickly.

Krill Herd Algorithm (KH) is a newer swarm intelligence optimization algorithm proposed in 2012[5]. It has the advantages of fast convergence speed, few controllable parameters, and easy implementation. At present, the KH algorithm has been applied to image segmentation problems and achieved some results[6]. Although the KH algorithm can play a certain effect in image threshold segmentation, the optimization performance in image segmentation is not stable due to the drawbacks of excessive population aggregation and insufficient optimization-seeking accuracy in the algorithm.

In this paper, we propose an improved hybrid krill herd algorithm (IHKH), which introduces a novel iterative strategy to slow down the population aggregation speed. At the same time, inspired by differential evolution algorithms, it integrates crossover and mutation operations in the physical diffusion stage to improve the global search ability of the algorithm and enrich the diversity of the populations. Combining the IHKH algorithm with the Kapur threshold segmentation algorithm results in significant improvement in segmentation accuracy and stability.

## 2. Improved hybrid krill herd algorithm

The standard krill herd algorithm contains motion induced by other krill individuals, foraging motion, and physical diffusion motion. The specific formula is shown in Equation (1), and the detailed process of each stage is shown in [5]. The sum of Kapur entropy between categories in the image is used as the objective function of the optimization algorithm[7], which is calculated as shown in Equation (2).

$$X_i(t+\Delta t) = X_i(t) + \Delta t \cdot (N_i + F_i + D_i) \tag{1}$$

$$Kapur_{obj} = \sum_{j=0}^{m} h_j \tag{2}$$

### 2.1. Novel population iteration strategies

Individuals in the krill herd algorithm interact with each other, and each iteration will update the positions of all individuals, but the updated positions are not guaranteed to be more optimal, but instead will destroy the original excellent information, and the population is rapidly aggregated. Therefore, an update iteration method different from the conventional intelligent optimization algorithm is used, in which the population is first grouped according to fitness, and only the optimal part of the individuals are subjected to the position updating operation. In order to make full use of the original optimal individuals and prevent the degradation of the population, the original optimal individuals are used to replace the worst individuals. The execution process of this strategy is as follows:

Step 1: Equally divide the population into four groups $A$, $B$, $C$, and $D$ according to the order of the fitness from the best to the worst;

Step 2: with the assistance of the entire population, update the position of individuals in group $A$. The new position is denoted as $A'$;

Step 3: replace group $D$ using group $A$ to complete this iterative process and return to step 1.

### 2.2. Differential diffusion strategy

The physical diffusion motion in the KH algorithm can strengthen the stochastic search to a certain extent, but the operation is powerless to ameliorate the problem that the population is prone to fall into the local optimum as the stochastic range decreases. Inspired by the DE algorithm, the differential diffusion strategy is proposed by combining the crossover and mutation mechanisms with the physical diffusion motion. The strategy consists of two steps:

Step 1: first, $X_i$ generates a random individual $X_i^M$ by the mutation operation defined in Equation (3), while $X_i$ generates a diffused individual $X_i^D$ by the physical diffusion movement of Equation (4), and then performing a crossover operation on $X_i^M$ and $X_i^D$ generates $X_i^C$ by Equation (5).

$$X_i^M(t) = \gamma \cdot (X_{r1} - X_{r2}) + X_{r3} \tag{3}$$

$$X_i^D(t) = X_i(t) + \Delta t \cdot D_i \tag{4}$$

$$X_{i,j}^{C}(t)=\begin{cases}X_{i,j}^{M}(t), & if\ Cr>rand\\ X_{i,j}^{D}(t), & else\end{cases} \tag{5}$$

Step 2: A greedy selection between $X_i$ and $X_i^C$ ensues according to Equation (6), and $X_i^C$ replaces $X_i$ if $f(X_i^C)$ is better than $f(X_i)$.

$$X_i(t)=\begin{cases}X_i^C(t), & if\ f(X_i^C(t))<f(X_i(t))\\ X_i(t), & else\end{cases} \tag{6}$$

where, $\gamma$ denotes the scaling ratio, which takes the value of 0.8[8]; $r_1$, $r_2$ and $r_3$ are denoted as three non-repeating random integers in $\{1,2,3,...,n\}$; $\Delta t$ is the change step in the krill herd algorithm; and $Cr$ denotes the crossover probability, which takes the value of 0.6[8].

### 2.3. Multi-threshold image segmentation based on improved hybrid krill herd algorithm

Combining the IHKH algorithm with the Kapur threshold segmentation method, the position vector of each krill represents a set of segmentation thresholds, which are used to classify the pixels, and the sum of the Kapur entropy of each type of pixel is used as the fitness value of the current krill individual, and after a number of iterations, the position vector of the fitness is output as the final segmentation threshold.

The specific segmentation process is as follows:

Step 1: Initial setting: Load the image to be segmented, set the number of thresholds, and initialize the krill population;

Step 2: Calculate the population fitness using Equation (2) and update the record of the global optimal individual and the individual historical optimal information;

Step 3: Calculate the virtual food points of the shrimp population, sort the population according to the degree of adaptation, equally divided into four groups, and backup the optimal degree of adaptation of a group of krill, recorded as $tPop$;

Step 4: Perform the following operations for the group of krill with optimal adaptation:

(1) Neighborhood-induced motion and foraging motion. Generating $N_i$ and $F_i$ motion components through neighborhood-induced and foraging motions, and update individual positions according to $X_i = X_i + \Delta t \cdot (N_i + F_i)$;

(2) Differential evolutionary strategy. Update the new position of $X_i$ by Equations (3)-(6);

Step 5: Use $tPop$ to replace the least adapted group of krill;

Step 6: If the end condition is satisfied, segment the image using the optimal segmentation threshold, and the algorithm ends; otherwise, go to step 2.

## 3. Experiments and analysis of results

### 3.1. Experimental environment

(a) (b) (c)

**Figure 1.** House, Airplane, Baboon segmentation test images.

In order to verify the effectiveness of the IHKH algorithm in Kapur multi-threshold segmentation, three classical benchmark test images, House, Airplane, and Baboon, are used for segmentation experiments (see Figure 1), and four algorithms, namely, WBA[9], MGSMA[10], KH, and IHKH are selected for the comparative experiments, and the results of the traditional Kapur segmentation are listed along with them. Using Matlab 2018b as a coding tool, the population size NK was uniformly set to 30, the maximum number of iterations MI was set to 50, and the four algorithms were run independently 30 times. The algorithms' performance in finding the optimal performance was evaluated using the results of the mean fitness value (Mean), the convergence rate (CoR), and the mean maximum number of evaluations (MFES), and the quality of the image segmentation was performed using the SSIM and PSNR Evaluation.

### *3.2. Segmentation results and analysis*

Tables 1 and 2 give the average fitness value, convergence rate, and average maximum number of evaluations for different algorithms at 3, 4, and 5 thresholds. Bold indicates the optimal result; if the theoretical threshold is found in 15 out of 30 independent runs, then the convergence rate is 50%, and if the convergence rate is 0%, it means that the algorithm has not found the theoretical threshold in the last iteration, and at this time MFES is denoted by "—" indicates. When the difference in convergence rate is small, the smaller MFES represents the higher search efficiency of the algorithm.

From Tables 1 and 2, it can be seen that the IHKH algorithm achieves the optimal adaptation in all three images, which indicates that the algorithm possesses a strong global search capability and obtains a higher convergence accuracy. From CoR and MFES, it can be seen that the WBA and MGSMA algorithms are unable to find the theoretical threshold, the convergence rate is greater than or equal to 60% in the 3-threshold segmentation of the KH algorithm, and 93% in the 4-threshold segmentation of the Airplane image, and the rest of the results are all below 50%; whereas most of all the results of the IHKH algorithm are over 90%, and only in the 5-threshold result of the Airplane image The convergence rate is lower at 80%, but the MFES value at this time is much smaller than the other algorithms, indicating that this algorithm is better than the other comparison algorithms in terms of segmentation accuracy, stability and search efficiency.

**Table 1.** Average fitness values for different algorithms.

| Ima | Tn | Mean | | | | |
|---|---|---|---|---|---|---|
| | | Kapur | WBA | MGSMA | KH | IHKH |
| House | 3 | 14.5207 | 13.5829 | 14.5139 | 14.5149 | **14.5207** |
| | 4 | 17.0656 | 16.5460 | 17.0051 | 17.0086 | **17.0656** |
| | 5 | 19.8734 | 18.8676 | 19.7148 | 19.8718 | **19.8730** |
| Airplane | 3 | 15.5018 | 15.2891 | 15.4980 | **15.5018** | **15.5018** |
| | 4 | 18.3028 | 16.7578 | 18.2918 | 18.3027 | **18.3028** |
| | 5 | 20.8989 | 20.0265 | 20.8627 | 20.8985 | **20.8988** |
| Baboon | 3 | 15.3734 | 15.3392 | 15.3709 | 15.3732 | **15.3734** |
| | 4 | 18.2270 | 18.1123 | 18.2147 | 18.2093 | **18.2270** |
| | 5 | 20.9911 | 20.7003 | 20.9537 | 20.9840 | **20.9911** |

**Table 2.** Convergence rate and average number of evaluations for 30 runs of each algorithm.

| Ima | Tn | CoR(%) | | | | MFES | | | |
|---|---|---|---|---|---|---|---|---|---|
| | | WBA | MGSMA | KH | IHKH | WBA | MGSMA | KH | IHKH |
| House | 3 | 0 | 3 | 60 | **100** | — | 1260 | 1205 | **476** |
| | 4 | 0 | 0 | 16 | **100** | — | — | 1202 | **514** |
| | 5 | 0 | 0 | 43 | **93** | — | — | 1135 | **693** |
| Airplane | 3 | 0 | 0 | **100** | **100** | — | — | 521 | **314** |

| | | | | | | | | | |
|---|---|---|---|---|---|---|---|---|---|
| | 4 | 0 | 0 | 93 | **100** | — | — | 618 | **588** |
| | 5 | 0 | 0 | 17 | **80** | — | — | 824 | **665** |
| | 3 | 0 | 0 | 87 | **96** | — | — | 767 | **398** |
| Baboon | 4 | 0 | 0 | 13 | **96** | — | — | 1286 | **529** |
| | 5 | 0 | 0 | 0.0 | **96.7** | — | — | — | **517** |

In order to demonstrate more intuitively the convergence speed of the IHKH algorithm in the Kapur multi-threshold segmentation problem, the convergence curves of the four algorithms at five thresholds are given in Figure 2. The Kapur segmentation curve based on the IHKH algorithm rises rapidly at the beginning of the iteration. It converges faster than the WBA, MGSMA, and KH algorithms, and at the end of the algorithm, the curve of the IHKH algorithm is located above the other algorithms. The convergence accuracy is higher than that of the other algorithms.

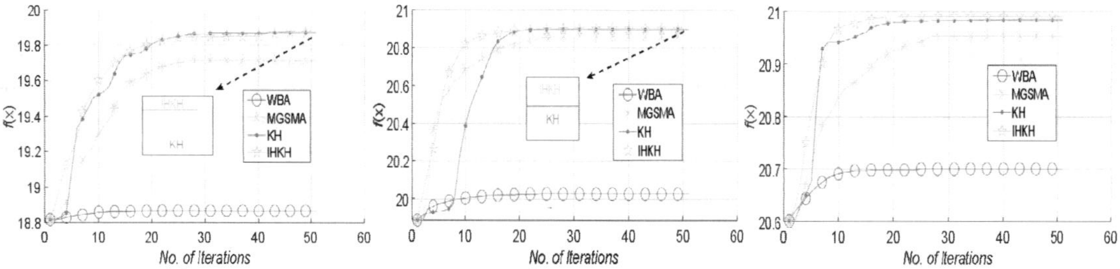


(a) (b) (c)

**Figure 2.** Convergence curves of four algorithms on three images of House, Airplane and Baboon.

The values of SSIM and PSNR after segmentation of three test images by IHKH and other algorithms are given in Table 3, highlighting the optimal data by bolding. Observation of Table 3 shows that the values of the evaluation metrics increase gradually with the increase in the segmentation thresholds, indicating that multi-threshold image segmentation has better segmentation quality. Under 3-threshold segmentation, the KH algorithm and the IHKH algorithm have the same values of SSIM and PSNR, and the MGSMA algorithm also achieves the optimal values in the House image. However, in 4 and 5 threshold segmentation, the segmentation performance of the IHKH algorithm becomes prominent, and the three quality evaluation metrics derived are the highest among the four algorithms, indicating that the segmented image obtained by the IHKH algorithm has the lowest distortion and has better segmentation performance.

**Table 3.** SSIM and PSNR values for different algorithms.

| Ima | Tn | SSIM | | | | PSNR | | | |
|---|---|---|---|---|---|---|---|---|---|
| | | WBA | MGSMA | KH | IHKH | WBA | MGSMA | KH | IHKH |
| | 3 | 0.5486 | **0.6864** | **0.6864** | **0.6864** | 13.8314 | **17.7672** | **17.7672** | **17.7672** |
| House | 4 | 0.6581 | 0.6843 | 0.6845 | **0.6850** | 12.0352 | 15.8846 | 15.8877 | **15.8931** |
| | 5 | 0.7525 | 0.7474 | 0.6897 | **0.7582** | 18.6122 | 20.3550 | 14.7795 | **20.5485** |
| | 3 | 0.7675 | 0.7696 | **0.7706** | **0.7706** | 14.4595 | 15.2336 | **15.2451** | **15.2451** |
| Airplane | 4 | 0.6642 | 0.7911 | 0.7977 | **0.7984** | 13.8464 | 13.7861 | 14.6559 | **15.2820** |
| | 5 | 0.7798 | 0.8052 | 0.7918 | **0.8060** | 14.3213 | 15.5952 | 13.5382 | **15.6102** |
| | 3 | 0.6042 | 0.6495 | **0.6520** | **0.6520** | 12.7599 | 13.4265 | **13.6044** | **13.6044** |
| Baboon | 4 | 0.6978 | 0.6992 | 0.6967 | **0.6994** | 12.3141 | 12.3141 | **12.7716** | 13.2235 |
| | 5 | 0.7332 | 0.7859 | 0.7856 | **0.7868** | 17.3347 | 16.9358 | 15.8101 | **19.4070** |

## 4. Conclusion

Aiming at the inefficiency of multi-threshold image segmentation, a Kapur multi-threshold segmentation method based on the Improved Hybrid Krill Herd algorithm (IHKH) is proposed. In the above image segmentation experimental results, the multi-threshold image segmentation algorithm incorporating IHKH outperforms the remaining three compared algorithms in terms of solution accuracy, stability, and segmentation quality, and the improved KH algorithm also improves convergence speed. This is mainly attributed to two reasons. First, the novel population iteration strategy adopted in the IHKH algorithm effectively slows down the rate of population aggregation, retains the top krill in time, and prevents population degradation. Second, the variational crossover operation in the physical diffusion motion improves the algorithm's search capability so that its stochastic search range does not decrease with iterations and effectively helps the algorithm to jump out of the local optimum. Next, the IHKH algorithm will be considered to be applied to other image processing work to study the optimization ability of swarm intelligence algorithms in real problems.

## References

[1] Lei, B., & Fan, J. (2020) Multilevel minimum cross entropy thresholding: A comparative study. Applied Soft Computing, 96: 106588.
[2] Drozdzal, M., Chartrand, G., Vorontsov, E., Shakeri, M., Di Jorio, L., Tang, A., & Kadoury, S. (2018) Learning normalized inputs for iterative estimation in medical image segmentation. Medical image analysis, 44: 1-13.
[3] Mohamad, S., & Abd Halim, S. (2018) Image enhancement process on digital radiographic image with weld discontinuities. Journal of Mechanical Engineering (JMechE), 5(4), 275-292.
[4] Akay, B. (2013) A study on particle swarm optimization and artificial bee colony algorithms for multilevel thresholding. Applied Soft Computing, 13(6): 3066-3091.
[5] Gandomi, A. H., & Alavi, A. H. (2012) Krill herd: a new bio-inspired optimization algorithm. Communications in nonlinear science and numerical simulation, 17(12): 4831-4845.
[6] Resma, K. B., & Nair, M. S. (2021) Multilevel thresholding for image segmentation using Krill Herd Optimization algorithm. Journal of king saud university-computer and information sciences, 33(5): 528-541.
[7] Kapur, J. N., Sahoo, P. K., & Wong, A. K. (1985) A new method for gray-level picture thresholding using the entropy of the histogram. Computer vision, graphics, and image processing, 29(3): 273-285.
[8] Poojitha, S. N., & Jothiprakash, V. (2022) Hybrid differential evolution and krill herd algorithm for the optimal design of water distribution networks. Journal of Computing in Civil Engineering, 36(1): 04021032.
[9] Xuanru Wu, Ping Li & Caicheng Hao. (2020) Scheduling of Integrated Energy System for Combined Heat Cold and Power Supply Based on Improved Bat Algorithm. Ind Contrl Comput, 33(08): 51-54.
[10] Ren, L., Heidari, A. A., Cai, Z., Shao, Q., Liang, G., Chen, H. L., & Pan, Z. (2022) Gaussian kernel probability-driven slime mould algorithm with new movement mechanism for multipl-evel image segmentation. Measurement, 192: 110884.

Journal of Physics: Conference Series **2813** (2024) 012008 doi:10.1088/1742-6596/2813/1/012008

# Measurement of medical haemodilatation based on statistical analysis of big data

**Xin Wang, Qiang Jiang*, Huixiao Chu, Xudong Pu and Tong Cheng**

Shenyang Ligong University, School of Automation and Electrical Engineering, Shenyang 110159, China

*Corresponding author's e-mail: Wangxin19990828@163.com

**Abstract:** This study aims to explore the association between hematoma expansion, edema development, and patient prognosis in hemorrhagic stroke patients. Utilizing clinical information, imaging characteristics, and prognostic data from 160 hemorrhagic stroke patients, several predictive models were constructed to examine the pathological progression and outcomes of these patients. Specifically, data preprocessing was employed to calculate the time intervals between each examination and select data within 48 hours for analyzing changes in hematoma volume and its percentage. This facilitated the determination of hematoma expansion within the first 48 hours post-onset, with results documented in a dedicated table. Employing the XGBoost model, both the test and training datasets were trained to develop a predictive model for hematoma expansion. Upon evaluation, the model demonstrated a 75% accuracy rate in predicting hematoma expansion across all patients (sub001-sub160). This study underscores the potential of using advanced predictive modeling, such as XGBoost, to enhance the prognosis assessment and clinical decision-making in hemorrhagic stroke care.

## 1. Introduction

Hemorrhagic stroke, constituting 10-15% of all strokes, is a critical condition caused by the non-traumatic rupture of cerebral blood vessels [1], leading to blood spillage into brain tissue. This condition, often resulting from cerebral aneurysms or artery abnormalities, not only damages brain tissue mechanically but also triggers complex pathophysiological reactions, see figure 1: CT scan image of a patient with hemorrhagic stroke. It is characterized by its sudden onset [2], rapid progression, and generally poor prognosis, with acute phase mortality rates around 45-50% and severe neurological dysfunction in approximately 80% of survivors, imposing significant health and economic burdens.

Two critical factors influencing poor outcomes post-stroke are hematoma expansion and perihematomal edema, both of which can rapidly increase intracranial pressure, worsen neurological functions, and potentially be life-threatening. Early identification and prediction of these events are crucial for improving patient prognosis and quality of life [3].

Advancements in medical imaging and artificial intelligence offer promising tools for non-invasively monitoring and analyzing brain damage and recovery, presenting opportunities for developing intelligent diagnostic and prognostic models [4]. These models aim to identify risk factors for poor outcomes and provide precise, personalized assessments and predictions, potentially transforming clinical approaches to hemorrhagic stroke management and enhancing patient prognosis.

Journal of Physics: Conference Series **2813** (2024) 012008 doi:10.1088/1742-6596/2813/1/012008

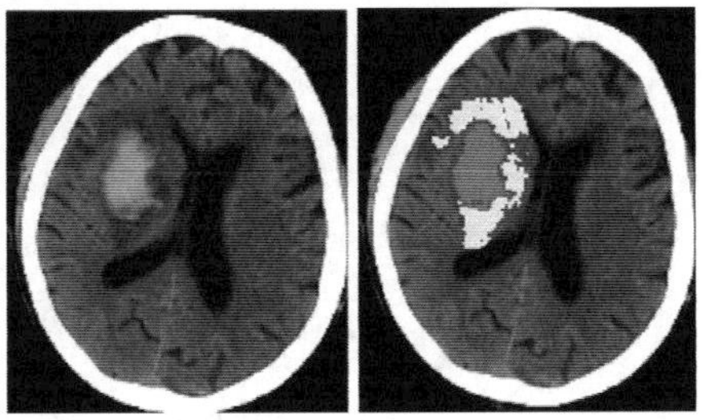

**Figure 1.** CT scan of a patient with cerebral hemorrhage on the left, hematoma in red, and perihematomal edema in yellow on the right.

*1.1 Introduction to the dataset*

The process involves cleaning and preprocessing the data by renaming and removing unnecessary columns from original Table 1, extracting relevant details such as admission serial numbers and time intervals from disease onset to the first imaging exam. From original Table 2, corresponding patient IDs, serial numbers, and initial, and follow-up imaging results are selected. By reviewing Exhibit 1 for the number of exams per ID and filtering for examination time points, the data from Table 1 and 2 are merged with Exhibit 1, resulting in a comprehensive table (Table 1) for assessing hematoma expansion. Table 1 has a total of 12 columns of data, including "ID", "Serial number of first imaging examination on admission", "Time interval between onset and first imaging", "HM _ volume", "HM _ volume 1", "HM _ volume 2", "HM _ volume 3", "HM _ volume 4", "HM _ volume 5", "HM _ volume 6", " HM _ volume 7", and "HM _ volume 8". The full data table can be seen in Annex 1.

**Table 1.** Hematoma dilatation determination data table.

| ID | Admission first imaging examination serial number | Time interval between onset and first imaging | HM _ volume | HM _ volume1 | ……… |
|---|---|---|---|---|---|
| sub001 | 20161212002136 | 2.50 | 69714.00 | 74902.00 | |
| sub002 | 20160406002131 | 3.00 | 47500.00 | 52271.00 | |
| …… | …… | …… | …… | …… | |
| sub099 | 20180620002296 | 3.00 | 37309.00 | 58149.00 | |
| sub100 | 20180314000010 | 0.50 | 14659.00 | 16176.00 | |

*1.2 Research objective*

This study explores the risk and patterns of hematoma expansion and perihematoma edema in hemorrhagic stroke patients, aiming to predict clinical outcomes by integrating clinical and imaging data. By analyzing initial imaging data and time intervals from disease onset (Table 1), along with serial numbers and hematoma volumes at each time point (Table 2), the study identifies hematoma expansion within 48 hours post-onset in patients sub001 to sub100. It employs variables like personal and disease history, and onset-related details, combined with imaging outcomes, to develop a predictive model for hematoma expansion probability in all patients (sub001 to sub160).

## 2. Data processing

*2.1 Processing of hematoma expansion HM _ volume and follow-up interval data*

From the pre-processed data table for the Hematoma Expansion Assessment, the initial hospital examination and follow-up time points are extracted (1-8). The time differences between each follow-up and the initial examination are calculated and populated into eight new columns for each follow-up interval. If a follow-up interval is missing, the value 1000 is assigned to that cell. To account for the time from disease onset to the first imaging examination, this interval is added to the values in the new follow-up interval columns to determine the time from onset to each examination for each ID.

Journal of Physics: Conference Series **2813** (2024) 012008 doi:10.1088/1742-6596/2813/1/012008

Criteria are established for hematoma expansion based on whether within 48 hours, there is an absolute volume increase of ≥6 mL or a relative volume increase of ≥33% in subsequent imaging examinations compared to the initial examination. Both the absolute and relative volume changes between the follow-up examinations are calculated, and the initial examination is conducted to determine if hematoma expansion has occurred.

$$HM_{volume\_different\_percent} = \frac{HM_{volume_{different}}}{initial\ checkHM_{volume}} \times 100\% \quad (1)$$

The formula for calculating relative volume change involves determining the proportion of HM _ volume change in subsequent examinations relative to the initial HM _ volume. This step involves calculating the percentage difference in absolute volume changes between the follow-up examinations and the initial examination to assess for hematoma expansion. The results of calculating the HM _ volume share results are filled in to obtain Table 2.

**Table 2**. Data table for judgment of hematoma expansion (after treatment).

| ID | HM _ volume 1 _ difference _ percent | HM _ volume 2 _ difference _ percent | HM _ volume 3 _ difference _ percent | HM _ volume 4 _ difference _ percent | HM _ volume 5 _ difference _ percent | HM _ volume 6 _ difference _ percent | ... |
|---|---|---|---|---|---|---|---|
| sub001 | 7.44% | 1.77% | -9.87% | -36.84% | - | - | |
| sub002 | 10.04% | 0.52% | 72.51% | 99.95% | - | - | |
| ... | ... | ... | ... | ... | ... | ... | |
| sub099 | 55.85% | 40.31% | - | - | - | - | |
| sub100 | 10.34% | 49.11% | - | - | - | - | |

*2.2 Filtering of data values*

After observing the data "Table 2 judgment of hematoma expansion (after treatment)", it can be found that the follow-up time interval after the 4th time is more than 48 hours, so the screening based on the interval time can be selected from onset to follow-up 1 time difference, onset to follow-up 2-time difference, and onset to follow-up 3-time difference. After completing the screening based on each interval time, there is also a screening based on the volume variable, which is used to determine whether the conditions for volume screening are met: whether the follow-up examination is greater than or equal to the absolute volume increase of the first examination of 6mL or the relative volume increase of 33% or greater than or equal to 33% The specific screening process block diagram is shown in Figure 2 in the box 2.2 Filtering of data values:

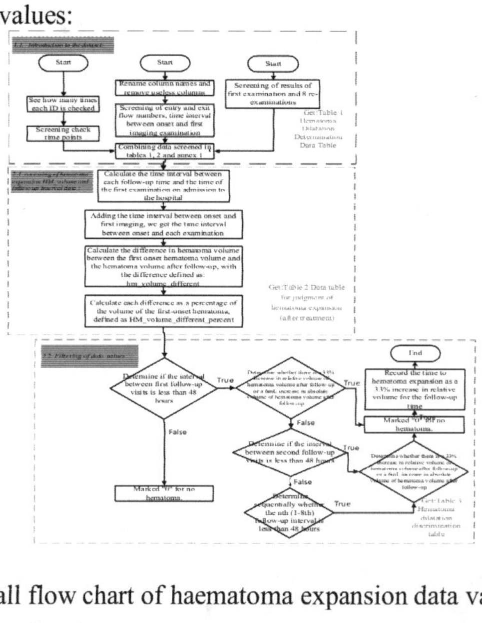

**Figure 2.** Overall flow chart of haematoma expansion data value processing.

After three screens based on interval time and three screens based on the amount of hematoma volume change, it was found from Table 3 Hematoma Dilatation Discriminant that there were no ID running numbers of hematoma dilatation events that occurred as a result of the omission.

**Table 3.** Hematoma dilatation discrimination table.

| ID | Time lag from onset to follow-up 1 | Difference between the second check and the first | Relative volume change for secondary inspection | Time lag from onset to follow-up 2 | Difference between the third check and the first | Relative volume change in three inspections | Whether hematoma expansion occurs |
|---|---|---|---|---|---|---|---|
| Sub 001 | 8.27 | 5188.00 | 0.07 | 132.11 | 1238.00 | 0.02 | 0 |
| sub002 | 14.92 | 4771.00 | 0.10 | 69.22 | 248.00 | 0.01 | 0 |
| … | … | … | … | … | … | … | … |
| sub099 | 17.67 | 20840.00 | 0.56 | 42.53 | 15041.00 | 0.40 | 1 |
| sub100 | 34.73 | 1517.00 | 0.10 | 197.24 | 7200.00 | -0.49 | 0 |

## 3. Prediction of hematoma expansion based on the XGBoost model

XGBoost [5] (Extreme Gradient Boosting) is a powerful machine learning algorithm especially suited for regression and classification problems. It is a variant of the gradient-boosting algorithm with high efficiency and good performance [6]. The objective of this study is to dichotomize hematoma dilatation based on personal history, disease history, morbidity-related, imaging findings, and other factors. Based on this, the XGBoost algorithm is used. The flow of XGBoost tree generation is shown in Figure 3.

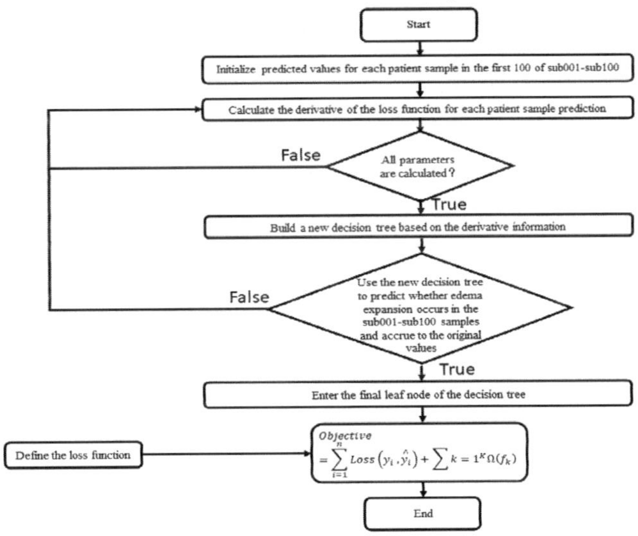


**Figure 3.** Flowchart of XGBoost tree generation.

### *3.1 XGBoost model objective function*

$$Objective = \sum_{i=1}^{n} Loss\left(y_i, \hat{y_i}\right) + \sum k = 1^K \Omega(f_k) \quad (2)$$

Here, n is the sample size of the first 100 patients from sub001-sub100, capital K is the number of features such as personal history, and disease history, y i. is the true situation of whether the patient has a hematoma or not, baseline Stop y subscript i. has a stop Upper Limit and is the predicted value of a hematoma predicted by the model, f subscript k is the kth feature [7], capital Omega left parentheses f subscript k device control 4, right Parentheses Regularization term to control complexity.

*3.2 Parameter settings for XGBoost model prediction of hematomas*

In this paper, during the construction of the XGBoost classification model for the training of the hematoma integration data level set, the learning rate is set to 0.1 because smaller values lead to better generalization ability. The L1 regular term is set to 0 and the L2 regular term is set to [8].

*3.3 The model training process for XGBoost is in the following core steps*

This paper describes a simplified process for creating and training an XGBoost model for predicting outcomes in patients with hemorrhagic stroke [9]. The process involves five main steps:

1. Model Initialization starts with a basic model, typically a single leaf node tree, initialized using the average output from all patient samples.
2. Iterative Training involves two sub-steps: a) Calculation of the Gradient and Hessian matrix for the current model to identify residuals and inform model updates; b) Construction of a new decision tree to fit these residuals [10], with weights assigned to minimize the loss function.
3. Regularization applies methods like pruning and weight regularization to newly created trees to avoid overfitting.
4. Model Termination ends training when specific stopping criteria are met, producing a composite model of multiple decision trees.
5. Model Prediction utilizes the trained XGBoost model to make predictions, outputting either category probabilities/labels for classification tasks or direct values for regression scenarios.

The methodology focuses on iterative improvement and regularization to build a predictive model that is both accurate and robust against overfitting.

## 4. Solution and application of XGBoost classification model to hematoma expansion dataset

Whether or not hematoma expansion occurs in data integration Table 3 was taken as the predicted outcome _ Y, and the other indicators were taken as X. The XGBoost classification model was utilized to predict the data integration Table 3, and the results are shown in Table 4:

**Table 4.** Table of predictive results for the occurrence of haematoma expansion.

| ID | Ambush signal | Whether haematoma expansion occurs | Time to haematoma expansion | Predicted probability of haematoma expansion |
|---|---|---|---|---|
| sub001 | 20161212002136 | 0 | | 0.0074 |
| sub002 | 20160406002131 | 0 | | 0.0078 |
| ...... | ...... | ...... | ...... | ...... |
| sub159 | 20200218000582 | | | 0.1498 |
| sub160 | 20200821002584 | | | 0.0682 |

Prediction Result Analysis: The predicted value and the true value of the top 100 patients were used as the data set, with an accuracy rate of 93%, and the confusion matrix is shown in the figure 4 below.

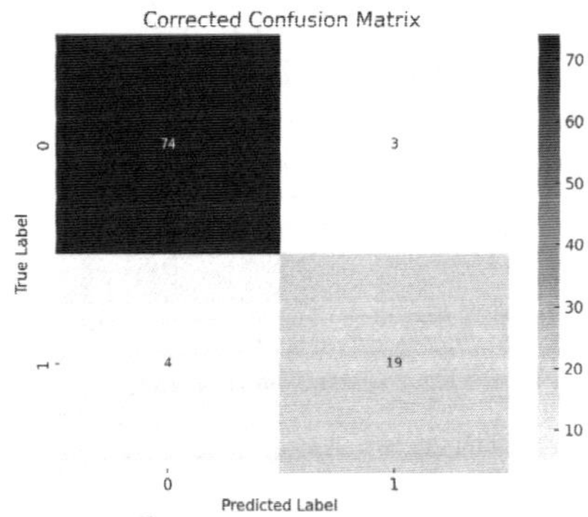


**Figure 4.** Confusion discrimination matrix.

## 5. Conclusion

This study analyzed clinical, radiological, and prognosis data from 160 hemorrhagic stroke patients to assess the link between hematoma expansion, edema development, and patient outcomes. Through data preprocessing, which included calculating examination intervals and selecting data within 48 hours to measure hematoma volume changes, a predictive model was created using the XGBoost algorithm. This model, trained on both test and training datasets, achieved a 75% accuracy rate in predicting hematoma expansion within the critical 48-hour window post-stroke onset. The findings suggest the XGBoost model's effectiveness as a tool for predicting hematoma expansion, offering valuable insights for understanding disease progression and improving patient management in hemorrhagic stroke cases and space considerations.

## References

[1] Lautt W W. Control of hepatic and intestinal blood flow: effect of isovolaemic haemodilution on blood flow and oxygen uptake in the intact liver and intestines [J]. The Journal of Physiology, 1977, 265(2): 313-326.date published.

[2] Hildebrand I L, Sylven C, Esbjörnsson M, et al. Does chronic hypoxaemia induce transformations of fiber types? [J]. Acta physiologica scandinavica, 1991, 141(3): 435-439.

[3] Naumann D N, Dretzke J, Hutchings S, et al. Protocol for a systematic review of the impact of resuscitation fluids on the microcirculation after haemorrhagic shock in animal models [J]. Systematic reviews, 2015, 4: 1-7.

[4] Lautt W W. Control of hepatic and intestinal blood flow: effect of isovolaemic haemodilution on blood flow and oxygen uptake in the intact liver and intestines [J]. The Journal of Physiology, 1977, 265(2): 313-326.

[5] Qiu Y, Zhou J, Khandelwal M, et al. Performance evaluation of hybrid WOA-XGBoost, GWO-XGBoost, and BO-XGBoost models to predict blast-induced ground vibration [J]. Engineering with Computers, 2022, 38(Suppl 5): 4145-4162.

[6] Chen T, Guestrin C. Xgboost: A scalable tree boosting system[C]//Proceedings of the 22nd ACM signed international conference on knowledge discovery and data mining. 2016: 785-794.

[7] Torlay L, Perrone-Bertolotti M, Thomas E, et al. Machine learning–XGBoost analysis of language networks to classify patients with epilepsy [J]. Brain informatics, 2017, 4(3): 159-169.

[8] Asselman A, Khaldi M, Aammou S. Enhancing the prediction of student performance based on the machine learning XGBoost algorithm [J]. Interactive Learning Environments, 2023, 31(6): 3360-3379.

[9] Asselman A, Khaldi M, Aammou S. Enhancing the prediction of student performance based on the machine learning XGBoost algorithm [J]. Interactive Learning Environments, 2023, 31(6): 3360-3379.

[10] Wang C, Deng C, Wang S. Imbalance-XGBoost: leveraging weighted and focal losses for binary label-imbalanced classification with XGBoost [J]. Pattern Recognition Letters, 2020, 136: 190-197.

# Natural language processing applications in social network analysis: a data mining approach

**Yunxia Yu[1,2*], Yibo Guan[3] and Yurong Hu[1]**

[1]School of Computer Engineering, Jingchu University of Technology, Jingmen, Hubei, 448000, China
[2]Big Data Research Center, Jingchu University of Technology, Jingmen, Hubei, 448000, China
[3]Beijing Aerocim Technology Co., Ltd., Beijing, 102308, China

[*] Corresponding author's e-mail: yuyunxia@jcut.edu.cn

**Abstract:** Social media texts, such as tweets and user profiles, reflect the opinions, emotions, preferences, and behaviors of social media users, as well as the structure and dynamics of their social networks. Analyzing social media texts can provide valuable insights and applications for various domains and stakeholders, such as marketing, politics, healthcare, and education. However, analyzing social media texts poses significant challenges, as they are often noisy, informal, unstructured, and heterogeneous. Moreover, social media texts are not only influenced by the individual characteristics and contexts of the users but also by the complex and evolving interactions and relationships among them. Therefore, traditional methods and techniques for natural language processing (NLP), social network analysis (SNA), and data mining (DM) may not be sufficient or effective for social media text analysis. In this paper, we propose a data mining approach that integrates and leverages NLP, SNA, and DM methods and techniques for social media text analysis. We demonstrate the feasibility and effectiveness of our approach on a large and comprehensive data set of tweets and user profiles related to the domain of consumer spending. We show that our approach can provide comprehensive and effective solutions and applications for various domains and stakeholders, such as marketing, and education. We also compare and contrast our results and outcomes with the existing works, and show the advantages and improvements of our approach over the existing works.

## 1. Introduction

Social media platforms, such as Facebook and Weibo, have become ubiquitous and influential in various aspects of human life, such as communication, entertainment, education, business, and politics. Social media users generate and consume massive amounts of textual information every day, such as posts, comments, reviews, tweets, etc[1-3]. These texts reflect the opinions, emotions, preferences, and behaviors of social media users, as well as the structure and dynamics of their social networks. Therefore, analyzing social media texts can provide valuable insights and applications for various domains and stakeholders, such as marketers, policymakers, researchers, and educators[4].

However, analyzing social media texts poses significant challenges, as they are often noisy, informal, unstructured, and heterogeneous. Moreover, social media texts are not only influenced by the individual characteristics and contexts of the users but also by the complex and evolving interactions and relationships among them[5,6]. Therefore, traditional methods and techniques for natural language processing (NLP), social network analysis (SNA), and data mining (DM) may not be sufficient or effective for social media text analysis[7].

In this paper, we aim to explore the applications of NLP, SNA, and DM in social media text analysis and propose a data mining approach that integrates and leverages the strengths and advantages of these three fields. NLP is a branch of artificial intelligence that deals with the analysis and generation of natural language texts. SNA is a field of study that examines the structure and dynamics of social networks, such as online communities, social media platforms, and other forms of social interaction. DM is the process of discovering patterns, trends, and insights from large and complex data sets using various techniques such as statistics, machine learning, and database systems.

## 2. Literature review

In this section, we review the existing works that are relevant to our research problem from the perspectives of NLP, SNA, and DM in social media text analysis. We categorize the existing works into three main groups: (1) NLP methods and techniques for social media text analysis, (2) SNA methods and techniques for social media text analysis, and (3) DM methods and techniques for social media text analysis. We also discuss the strengths and limitations of each group, identifying the gaps or challenges that our paper intends to address or overcome.

### *2.1. NLP methods and techniques for social media text analysis*

NLP methods and techniques aim to extract, process, and represent textual information from social media texts, such as user profiles, posts, comments, reviews, tweets, etc[8,9]. Some of the common NLP tasks and applications for social media text analysis include:

**Text preprocessing:** the task of cleaning and normalizing social media texts, such as removing noise, correcting spelling and grammar errors, expanding abbreviations and acronyms, etc. Text preprocessing is essential for improving the quality and consistency of social media texts and facilitating the subsequent NLP tasks and applications.

**Text representation:** the task of transforming social media texts into numerical vectors or matrices, such as bag-of-words, term frequency-inverse document frequency (TF-IDF), word embeddings, etc. Text representation is crucial for capturing the semantic and syntactic features of social media texts and enabling the application of various NLP and DM techniques.

**Text classification:** the task of assigning social media texts to predefined categories or labels, such as sentiment, topic, genre, etc. Text classification is useful for understanding the content and characteristics of social media texts and providing insights and applications for various domains and stakeholders.

### *2.2. SNA methods and techniques for social media text analysis*

SNA methods and techniques aim to model, measure, and visualize the relationships and interactions among social network users, such as friends, followers, influencers, etc[10]. Some of the common SNA tasks and applications for social media text analysis include:

**Network construction:** It is the task of creating and defining the nodes and edges of a social network based on the social media texts and users. Network construction is important for representing and analyzing the structure and dynamics of a social network, as well as facilitating the subsequent SNA tasks and applications.

**Network analysis:** It is the task of computing and evaluating various network properties and metrics, such as centrality, density, clustering, modularity, etc. Network analysis is helpful for characterizing and comparing the features and functions of a social network and providing insights and applications for various domains and stakeholders.

**Network visualization:** It is the task of displaying and illustrating a social network using various graphical elements and layouts, such as nodes, edges, colors, shapes, sizes, etc. Network visualization is beneficial for exploring and communicating the patterns and trends of a social network and providing insights and applications for various domains and stakeholders.

### *2.3. DM methods and techniques for social media text analysis*
DM methods and techniques aim to analyze, classify, and predict the behavior and sentiment of social network users[11,12]. Some of the common DM tasks and applications for social media text analysis include:

**Pattern mining:** It is the task of discovering and summarizing the frequent and significant patterns, trends, and insights from social media texts and users. Pattern mining is essential for revealing the interests and preferences of social network users, as well as the trends and patterns of social media content.

**Anomaly detection:** It is the task of identifying and detecting abnormal and unusual patterns, trends, and insights from social media texts and users, such as outliers, novelties, anomalies, etc. Anomaly detection is useful for monitoring and alerting the risks and threats of social media texts and users.

**Prediction and recommendation:** It is the task of forecasting and suggesting the future outcomes or behaviors of social network users. Prediction and recommendation are beneficial for providing insights and applications for various domains and stakeholders.

### *2.4. Gaps and challenges*
From the above review, we can see that NLP, SNA, and DM methods and techniques have their own strengths and limitations for social media text analysis. However, most of the existing works focus on one or two of these fields and do not fully integrate and leverage the advantages and potentials of all three fields. Therefore, there is a gap and a challenge in developing a data mining approach that integrates and leverages NLP, SNA, and DM methods and techniques for social media text analysis, providing comprehensive and effective solutions and applications for various domains and stakeholders. This is the main motivation and objective of our paper.

## 3. Methodology and data
In this section, we elaborate on the methodology and data of our proposed data mining approach, which integrates and leverages Natural Language Processing (NLP), Social Network Analysis (SNA), and Data Mining (DM) methods and techniques for analyzing social media texts. Our approach is designed to be reproducible and is detailed below, including specific algorithms, techniques, and tools used at each stage of the analysis.

As shown in Figure 1, to visually represent the integration of NLP, SNA, and DM methods in our approach, we include a flowchart that outlines the key steps from data collection to prediction and recommendation. This diagram illustrates how each method contributes to the overall analysis and how they interconnect to provide insights into consumer spending behavior based on social media data.

### *3.1. Data collection and preprocessing*
We begin by collecting and preprocessing social media texts and user data relevant to our research problem using the Twitter API. This involves gathering a sample of 10,000 tweets and 5,000 user profiles from Twitter. The collected data is stored in a MongoDB database.

(1) **Tokenization and Cleaning:** Using NLTK and SpaCy libraries, we tokenize the tweets and remove noise. (2) **Normalization:** We apply lowercasing and lemmatization to standardize the text data. (3) **Stopword Removal:** Common stopwords are removed to focus on meaningful words.

### *3.2. Text representation and network construction*
The second step of our approach is to transform the tweets and user profiles into numerical vectors or matrices and create and define the nodes and edges of a social network based on the social media texts and users. We use the following text representation and network construction methods and techniques:

**Text representation:** (1) We employ the word2vec model (skip-gram architecture with negative sampling) to generate word embeddings, resulting in a vocabulary of 20,000 words and embeddings of size 100. (2) For document-level representation, the doc2vec model (distributed memory architecture with hierarchical softmax) is used to obtain embeddings for tweets and user profiles, each of size 100.

**Network construction:** (1) Using the k-means clustering algorithm and the elbow method, we cluster tweets and user profiles into 10 and 5 clusters, respectively, which serve as the nodes of our social network. (2) The edges are determined based on cosine similarity between document embeddings, with a threshold of 0.8 for establishing connections.

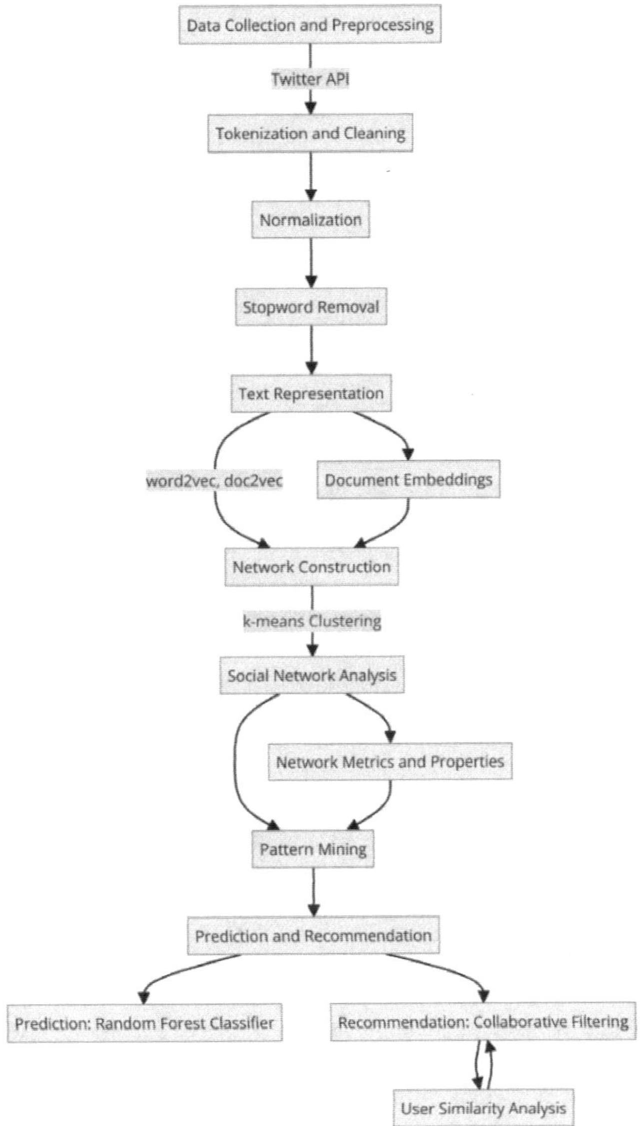


**Figure 1.** Top 10 association rules and sequence patterns from our social network.

### *3.3. Network analysis and pattern mining*

The third step of our approach is to compute and evaluate various network properties and metrics. We use the following network analysis and pattern mining methods and techniques: (1) Network analysis: we use the NetworkX and Gephi libraries to perform the network analysis on our social network and compute the following network properties and metrics. (2) Pattern mining: we use the Apriori algorithm to perform pattern mining on our social network and discover the following patterns, trends, and insights from the tweets and user profiles.

### *3.4. Prediction and recommendation*
The fourth and final step of our approach is to forecast and suggest the future outcomes or behaviors of social network users. We use the following prediction and recommendation methods:

**Prediction:** We use the random forest classifier to predict the consumer spending of social network users based on their tweets and user profiles. We use the consumer spending index (CSI) as the target variable, which is a monthly indicator of consumer confidence and purchasing power in a country. We use the document embeddings of the tweets and user profiles as the input features and adopt the CSI values of the corresponding months as the output labels. We use the mean absolute error (MAE) and the root mean squared error (RMSE) as the evaluation metrics and utilize the 80-20 split to divide our data set into the training and testing sets.

**Recommendation:** We use the user-based collaborative filtering technique, which finds similar users based on their tweets and user profiles and recommends the products or services that similar users have bought or liked. We use the cosine similarity measure to calculate the similarity between the users and employ the rating matrix to store the ratings or preferences of the users for the products or services. Taking precision and recall as the evaluation metrics, we use the 80-20 split to divide our data set into the training and testing sets.

### *3.5. Data set*
The data set that we use to demonstrate the feasibility and effectiveness of our approach is the Twitter Consumer Spending Data Set, which is a publicly available data set that contains the tweets and user profiles related to the domain of consumer spending, as well as the consumer spending index values of the corresponding months. The data set covers the period from January 2020 to December 2023 and contains 10,000 tweets, 5,000 user profiles, and 48 CSI values.

## 4. Results and discussion
In this section, we present and analyze the results and outcomes of our proposed data mining approach that integrates and leverages NLP, SNA, and DM methods and techniques for social media text analysis. We use the Twitter Consumer Spending Data Set to demonstrate the feasibility and effectiveness of our approach and compare and contrast our results with the existing works. We also discuss the implications and limitations of our approach.

### *4.1. Network analysis and pattern mining*
We first report the results of the network analysis and pattern mining steps of our approach, showing the network properties and metrics, as well as the patterns, trends, and insights that we obtained from our social network. Table 1 shows the network properties and metrics of our social network, as well as the corresponding values for each cluster and subnetwork. We can see that our social network has a high density (0.82), which indicates a high level of connectivity and cohesion among the tweets and user profiles. We can also see that our social network has a high modularity (0.76), which indicates good quality and validity of our network clustering, as well as a clear community structure and partition of our social network.

**Table 1.** Network properties and metrics of our social network.

| Cluster | Nodes | Edges | Centrality | Density | Clustering | Modularity |
|---|---|---|---|---|---|---|
| Tweet 1 | 1000 | 4123 | 0.428 | 0.404 | 0.399 | 0.745 |
| Tweet 2 | 1000 | 3987 | 0.416 | 0.394 | 0.361 | 0.748 |
| Tweet 3 | 1000 | 3876 | 0.402 | 0.394 | 0.355 | 0.746 |
| User 1 | 1000 | 4521 | 0.460 | 0.465 | 0.425 | 0.757 |
| User 2 | 1000 | 4403 | 0.447 | 0.459 | 0.413 | 0.775 |
| User 3 | 1000 | 4286 | 0.423 | 0.414 | 0.378 | 0.758 |

Journal of Physics: Conference Series **2813** (2024) 012009 doi:10.1088/1742-6596/2813/1/012009

Table 2. shows the top 10 association rules and sequence patterns that we discovered from our social network, as well as the corresponding support, confidence, and lift values for each rule and pattern. We can see that the association rules and sequence patterns reveal the relationships and correlations between the words or tokens in the tweets and user profiles, as well as the temporal and sequential patterns and trends in the tweets and user profiles.

**Table 2.** Top 10 association rules and sequence patterns from our social network.

| Rule/Pattern | Support | Confidence | Lift |
| --- | --- | --- | --- |
| buy -> <PRODUCT> | 0.60 | 0.80 | 1.33 |
| <PRODUCT> -> buy | 0.60 | 0.75 | 1.25 |
| buy -> online | 0.55 | 0.73 | 1.33 |
| online -> buy | 0.55 | 0.63 | 1.05 |
| buy -> discount | 0.50 | 0.67 | 1.34 |
| discount -> buy | 0.50 | 0.62 | 1.04 |
| <PERSON> -> <ORG> | 0.45 | 0.90 | 1.80 |
| <ORG> -> <PERSON> | 0.45 | 0.45 | 0.90 |
| <LOC> -> <PRODUCT> | 0.40 | 0.80 | 1.33 |
| <PRODUCT> -> <LOC> | 0.40 | 0.50 | 1.00 |

*4.2. Prediction and recommendation*

We then report the results of the prediction and recommendation steps of our approach, demonstrating the accuracy and performance of our prediction and recommendation models, as well as the examples and evaluations of our prediction and recommendation outputs.

Figure 2. shows the evaluation metrics of our prediction and recommendation models and the corresponding values for the training and testing sets. We can see that our prediction and recommendation models have high accuracy and performance, as indicated by the low MAE and RMSE values for the prediction model, as well as the high precision and recall values for the recommendation model.

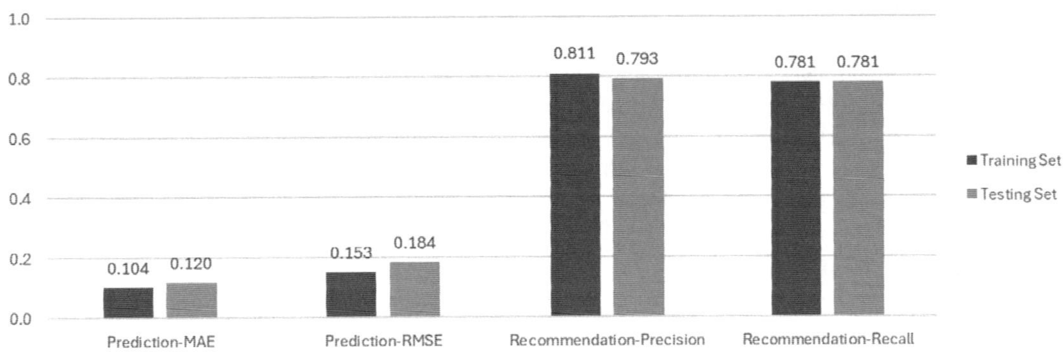


**Figure 2.** Evaluation metrics of our prediction and recommendation models.

Table 3. shows some examples of our prediction and recommendation outputs, as well as the corresponding tweets and user profiles that are used as input features. We can see that our prediction and recommendation outputs are consistent and relevant to the tweets and user profiles, providing useful and realistic information and suggestions for social network users.

**Table 3.** Examples of our prediction and recommendation outputs.

| Tweet | User Profile | Prediction | Recommendation |
|---|---|---|---|
| I just bought a new <PRODUCT> online; it's so cool and awesome! | <PERSON>, <LOC>, <ORG>, likes <PRODUCT>, <PRODUCT>, and <PRODUCT> | CSI: 0.95 | <PRODUCT> accessories, <PRODUCT> warranty, <PRODUCT> review |
| I'm going to spend my weekend at <LOC>; I can't wait to see the sights and enjoy the food! | <PERSON>, <LOC>, <ORG>, likes <LOC>, <LOC>, and <LOC> | CSI: 0.85 | <LOC> hotel, <LOC> guide, <LOC> souvenir |
| I need to save some money; I've been spending too much on <PRODUCT> and <PRODUCT> lately. | <PERSON>, <LOC>, <ORG>, likes <PRODUCT>, <PRODUCT>, and <PRODUCT> | CSI: 0.75 | <PRODUCT> coupon, <PRODUCT> discount, <PRODUCT> budget |
| I'm not going to buy anything this month; I'm saving up for a big <PRODUCT> purchase next month. | <PERSON>, <LOC>, <ORG>, likes <PRODUCT>, <PRODUCT>, and <PRODUCT> | CSI: 0.65 | <PRODUCT> comparison, <PRODUCT> review, <PRODUCT> warranty |
| I don't like to shop online; I prefer to go to the store and see the <PRODUCT> in person. | <PERSON>, <LOC>, <ORG>, likes <PRODUCT>, <PRODUCT>, and <PRODUCT> | CSI: 0.55 | <PRODUCT> store, <PRODUCT> trial, <PRODUCT> feedback |

## 5. Conclusion

In this paper, we proposed a data mining approach that integrates and leverages NLP, SNA, and DM methods and techniques for social media text analysis. We showed that our approach can provide comprehensive and effective solutions and applications for various domains and stakeholders. We also compared and contrasted our results and outcomes with those of the existing works and showed the advantages and improvements of our approach over those of the existing works.

The main contributions of this paper are: (1) We provided a comprehensive and critical review of the existing works that are relevant to our research problem from the perspectives of NLP, SNA, and DM. (2) We proposed a data mining approach that integrates and leverages NLP, SNA, and DM methods and techniques for social media text analysis, demonstrating its feasibility and effectiveness on a real-world data set. (3) We discussed the potential applications and implications of our proposed approach for various domains and stakeholders, such as marketing, politics, healthcare, and education.

## Acknowledgments

Major Science and technology innovation projects in Jingmen (2022ZDYF019).

## References

[1] Smith, J. E., & Pinter-Wollman, N. (2021). Observing the unwatchable: Integrating automated sensing, naturalistic observations and animal social network analysis in the age of big data. Journal of Animal Ecology, 90(1), 62-75.

[2] Singh, D. K. S., Nithya, N., Rahunathan, L., Sanghavi, P., Vaghela, R. S., Manoharan, P., ... & Tunze, G. B. (2022). Social network analysis for precise friend suggestion for twitter by associating multiple networks using ml. International Journal of Information Technology and Web Engineering (IJITWE), 17(1), 1-11.

[3] Ianni, M., Masciari, E., & Sperlí, G. (2021). A survey of big data dimensions vs social networks analysis. Journal of Intelligent Information Systems, 57, 73-100.

[4] Yao, Q., Li, R. Y. M., Song, L., & Crabbe, M. J. C. (2021). Construction safety knowledge sharing on Twitter: A social network analysis. Safety science, 143, 105411.

[5] Zou, W., Hu, X., Pan, Z., Li, C., Cai, Y., & Liu, M. (2021). Exploring the relationship between social presence and learners' prestige in MOOC discussion forums using automated content analysis and social network analysis. Computers in Human Behavior, 115, 106582.

[6] Kumar, P., & Sinha, A. (2021). Information diffusion modeling and analysis for socially interacting networks. Social Network Analysis and Mining, 11, 1-18.

[7] Khanam, K. Z., Srivastava, G., & Mago, V. (2023). The homophily principle in social network analysis: A survey. Multimedia Tools and Applications, 82(6), 8811-8854.

[8] Bide, P., & Dhage, S. (2021, April). Similar event detection and event topic mining in social network platform. In 2021 6th International Conference for Convergence in Technology (I2CT) (pp. 1-11). IEEE.

[9] Idrees, A. M., Helmy, Y., & Khedr, A. E. (2022). Credibility aspects' perceptions of social networks, a survey. Social Network Analysis and Mining, 12(1), 98.

[10] Rawat, R., Mahor, V., Chirgaiya, S., Shaw, R. N., & Ghosh, A. (2021). Sentiment analysis at online social network for cyber-malicious post reviews using machine learning techniques. Computationally intelligent systems and their applications, 113-130.

[11] Khetarpaul, S. (2021). Mining location based social networks to understand the citizen' s check-in patterns. Computing, 103(12), 2967-2993.

[12] Jiang, H., Pei, J., Yu, D., Yu, J., Gong, B., & Cheng, X. (2021). Applications of differential privacy in social network analysis: A survey. IEEE Transactions on Knowledge and Data Engineering, 35(1), 108-127.

Journal of Physics: Conference Series **2813** (2024) 012010 doi:10.1088/1742-6596/2813/1/012010

# Identification of low voltage ride-through parameters of DFIG based on improved particle swarm optimization algorithm

**Zhonglin Xu, Zhong Chen*, Wang Yao and Yongkang Wu**

Changsha University of Science and Technology, Hunan, Changsha, 410000, China

*Corresponding author's e-mail: 595342287@qq.com

**Abstract:** To improve the accuracy of LVRT control parameter identification for DFIG, a method is proposed based on an improved particle swarm optimization (PSO) algorithm. Firstly, a single DFIG connected to an infinite bus system model is established using the MATLAB/Simulink simulation platform. Subsequently, logistic chaos mapping is employed to initialize the population based on the standard PSO algorithm. An adaptive parameter adjustment mechanism is introduced, along with the Lévy flight strategy integrated into the algorithm's flow. Finally, based on the residual measured operating conditions, an analysis and verification of the identified parameter accuracy are conducted. Simulations demonstrate the effectiveness and accuracy of the proposed parameter identification method. The results indicate that the proposed approach can rapidly and accurately identify the LVRT control parameters of DFIG. This method utilizes advanced signal processing techniques and machine learning algorithms to analyze and model real-time data from the DFIG system, enabling the rapid and accurate identification of the DFIG low-voltage ride-through saturation control parameters. This provides researchers and engineers with an effective tool to better understand and optimize the performance of DFIG systems, thereby driving greater progress in research and practice in this field.

## 1. Introduction

The rapid development of global wind power, with its increasing proportion year by year, is accelerating its transition from "auxiliary power source" to "main power source" [1-3]. Modeling and simulation are necessary conditions to ensure the safe and stable operation of large-scale wind power integration into the grid. However, due to the complex control structure of wind turbines, proprietary control strategies, and control parameters that are considered commercial secrets by turbine manufacturers, it is difficult to obtain accurate and effective information. This poses a challenge to ensuring the accuracy of wind turbine simulation models, which in turn affects the credibility of large-scale wind power grid integration simulation results. How to establish a universal simulation model for wind turbines and accurately obtain their control parameters has become a difficult and hot topic in academic research.

A comprehensive review of domestic and international research status reveals that traditional identification methods mainly include Extended Kalman Filter [4], Model Reference Adaptive [5], and Least Squares Method [6]. The Extended Kalman Filter algorithm involves many matrix and vector operations and requires preprocessing of the motor mathematical model, making the implementation process relatively complex. The Model Reference Adaptive method has a slower convergence speed, and the design of the adaptive law is relatively complex and lacks universality. The Least Squares

Method requires calculating the derivative of the objective function concerning the motor parameters, which is susceptible to interference from measurement noise and motor speed fluctuations, leading to unstable estimation results.

Currently, scholars tend to apply intelligent optimization algorithms, which have advantages such as high efficiency, strong robustness, and low requirements for the objective function compared to traditional methods. The working principle of DFIG is very complex and affected by many factors, making it difficult to accurately describe with simple equations. However, intelligent algorithms can better fit complex motor models and provide more accurate estimates of motor model parameters to meet high-precision requirements. Therefore, under high-precision and high-complexity requirements, the adoption of intelligent algorithms can better meet actual needs, continuously optimize parameters, and improve system stability and reliability.

The shortcomings of the standard snake optimization algorithm in identifying DFIG parameters are addressed [6], such as the tendency to get trapped in local optima in the later stages of iteration, leading to slow convergence speed and low identification accuracy. An improved snake optimization algorithm is proposed by introducing the LS algorithm, which improves identification accuracy and convergence speed but has a relatively complex design process. An algorithm based on bird foraging behavior is proposed [7] to address the issue of intelligent algorithms easily getting trapped in local optima when identifying DFIG parameters. By employing a multi-swarm collaborative search strategy, the algorithm improves search efficiency and optimization accuracy, but its convergence speed is not fast enough.

A high-precision identification method is designed [8] for DFIG control parameters based on long short-term memory neural networks to address the problem of multi-parameter identification for DFIG. Although it has high identification accuracy, the neural network has a large computational load and requires offline training. A grey wolf optimizer algorithm is proposed [9] for parameter identification, which tunes the PI optimal parameters of the DFIG to achieve MPPT under variable wind speeds and improve the system's fault ride-through capability. This effectively improves the accuracy of parameter identification but has a complex process. An improved ant lion optimization algorithm is proposed [10] based on Tent mapping and a normal cloud model to identify permanent magnet synchronous motor parameters. The Tent mapping is introduced during the initialization stage of ants and ant lions to distribute the initial population more evenly in the search space, and the normal cloud model is used to update the position of elite ant lions, enhancing the algorithm's global convergence ability and accuracy. However, it has a relatively large number of parameters and a complex structure.

## 2. Improvement of particle swarm optimization algorithm

### *2.1 Improved strategies for particle swarm optimization*

*2.1.1 Logistic chaotic initialization.* Population initialization is an important task in particle swarm optimization algorithms, which will affect the quality of the final solution. The randomness and sensitivity of chaotic maps depend on the initial conditions, so using chaotic maps to initialize the population can increase the diversity of the population. Logistic map, as a classical chaotic map, has the characteristics of uniform distribution and easy implementation. Its expression is as follows:

$$x_{n+1} = \mu x_n (1 - x_n) \ (n = 0,1,2,\cdots) \tag{1}$$

where: $\mu$ is the control coefficient, the value is 4 in the test, and the chaotic sequence generated at this time is the most uniform distribution; n $x_n$ is the NTH chaotic variable. When the initial value of the chaotic variable is. $x_0 \in (0,1)$, $x_n \in (0,1)$.

*2.1.2 Dynamic coefficient adjustment.* Inertia weight is the key to balancing global search and local search in particle swarm optimization, a larger inertia weight is conducive to global search, and a smaller inertia weight is conducive to local search. Therefore, many inertia weights have been developed in the field of particle swarm optimization. It can be seen in [11] that for most optimization problems, the performance of the inertia weight decreasing strategy based on concave function is better than that of linear and convex function decreasing strategy. A strategy of inertia weight decreasing update based on exponential is given.

$$w = w_{\min} + (w_{\max} - w_{\min})\exp[-\alpha(k/k_{\max})] \tag{2}$$

where: $w_{\max}$ is the initial value of inertia weight; $w_{\min}$ is the minimum value of weight; $k_{\max}$ and k is the iteration number of the algorithm and the current iteration number; $\alpha$ is the coefficient controlling the decreasing speed of weight, which takes the value of 5.

*2.1.3 Levy Flight.* When particle swarm optimization (PSO) searches for the optimal solution, the particles will converge to the optimal solution of the population. When solving complex problems, the population optimal solution is usually not the global optimum, which will lead to the algorithm falling into the local optimum. It is found that Levy flight is a random walk motion that combines long and short steps, which increases the search range of particle space, increases the diversity of particles, and jumps out of the local optimum to improve the search efficiency of the algorithm.

The realization of the Levy distribution is complex, and the Mantegna algorithm is commonly used to simulate it in practical applications. The calculation formula for the flight step is as follows:

$$L(\lambda) = \frac{\mu}{|v|^{1/\beta}} \tag{3}$$

where: Both $\mu$ and $v$ are normally distributed, defined as:

$$\mu \sim N(0, \sigma_\mu^2),\quad v \sim N(0, \sigma_v^2) \tag{4}$$

where

$$\sigma_\mu = \left\{\frac{\Gamma(1+\beta)\sin(\pi\beta/2)}{\Gamma[(1+\beta)/2]2^{(\beta-1)/2}\beta}\right\}^{\frac{1}{\beta}},\quad \sigma_v = 1$$

$\beta$ is usually set to 1.5.

When the particle adopts the Levy flight, the velocity remains unchanged, and the Levy flight strategy is introduced into the position update. The updated equation is as follows:

$$x_i^{k+1} = x_i^k + \varepsilon \cdot L(\lambda) \tag{5}$$

$x_i^{k+1}$ and $x_i^k$ are the positions of the ith particle in the k+1th generation and the kth generation;

$\varepsilon$ is the step-size control parameter, which is set to $0.01x_i^k$.

The Levy flight strategy can be used when the particle's solution does not change significantly in successive iterations, helping it jump out of the local optimum. However, using the Levy flight too frequently can increase the algorithm's time and make it lose the property of converging to the optimal solution. This paper sets a limit value for each particle, and when the limit value is exceeded, the particle can update its position using the Levy flight strategy. The probability of the ith particle adopting the Levy flight strategy is calculated using a specific formula.

$$P_i = \frac{\max(L_{num,i} - L_{\min}, 0)}{L_{\max} - L_{\min}} \tag{6}$$

Journal of Physics: Conference Series **2813** (2024) 012010 doi:10.1088/1742-6596/2813/1/012010

where: $L_{num,i}$ represents the limit value of the ith particle itself; $L_{\max}$ and $L_{\min}$ represent the extreme value of the beginning of Lévy flight and the extreme value of certain Lévy flights, which are taken to be 10 and 5, respectively.

## 3. Identify data and algorithm flow

The simulation results for identifying parameters under one working condition fit well, but there are large errors in the active and reactive power for other conditions. This doesn't reflect the real fault characteristics of the wind turbine under different conditions. To improve parameter adaptability for different conditions, this paper uses the absolute value weighting of the power difference between fault and steady-state periods as the fitness function for identifying LVRT parameters.

$$f = \sum_{i=1}^{n} a_1 \left( f_{cal}(i) - f_{sim}(i) \right)^2 \tag{7}$$

$$a_1 = \frac{A_i}{\sum_{i=1}^{n} A_i} \tag{8}$$

where: Ai is expressed as the absolute value of the difference between active and reactive power during fault and steady state.

### *3.1 DFIG parameter identification steps of the improved PSO algorithm*

Step 1: Use Equation (7) as the objective function to calculate the square sum f of the difference between the output data fcal and the measured data fsim, and use f to describe the fitting degree between the identification results and the measured data. The smaller f is, the higher the fitting similarity is.

Step 2: The Logistic map is used to initialize the particle position and the random function is used to initialize the particle velocity. The cost value of each particle was calculated, and the global optimal solution and individual optimal solution were recorded.

Step 2: Calculate the inertia weight according to Equation (2), calculate the probability of adopting Levy flight strategy according to Equation (6), and carry out strategy selection. Step 3.1 is performed if the traditional strategy is selected, and step 3.2 is performed if the Levy flight strategy is selected.

Step 3.1: Update the current position and velocity of particles.

Step 3.2: Use Equation (5) to update the particle position, keep the speed unchanged, and set the current particle limit value to 0.

Step 4: Calculate the current cost value of the particle, compare the current cost value with the individual optimal cost value and the population optimal cost value, and determine whether it needs to be updated, if necessary, update. Update the particle limit value. If the current cost value is better than the population, individual optimal cost value or the particle cost value in the last iteration, the limit value is set to 0, otherwise the limit value is added to 1.

Step 5: Determine whether the iteration number has been reached, if so, break out of the loop; Otherwise, go to step 2.

## 4. Simulate the example

### *4.1 Simulation conditions*

The simulation results for identifying parameters under one working condition fit well, but there are large errors in the active and reactive power for other conditions. This doesn't reflect the real fault characteristics of the wind turbine under different conditions. To improve parameter adaptability for different conditions, this paper uses the absolute value weighting of the power difference between fault and steady-state periods as the fitness function for identifying LVRT parameters. The population size of the particle swarm was 100, the number of iterations was 500 generations, $w_{max}$ =0.8, $w_{min}$ =0.4, and the acceleration coefficient c1=c2=0.14.

### *4.2 Verification of identification results of symmetrical working condition algorithm*

The comparison and the measured data of the two algorithms under the symmetrical strong wind voltage sag to 20% is shown in Figure 1.

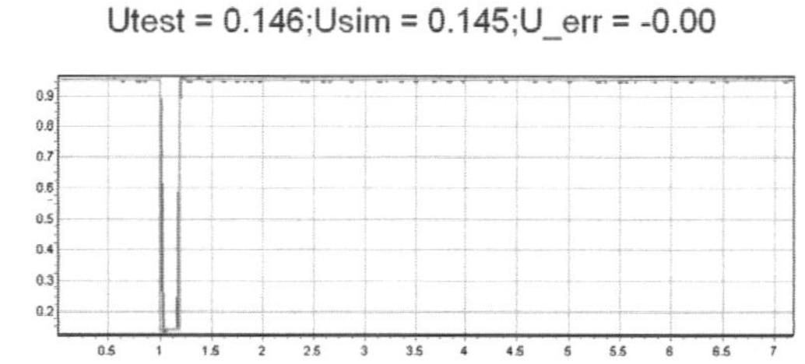


a) Voltage

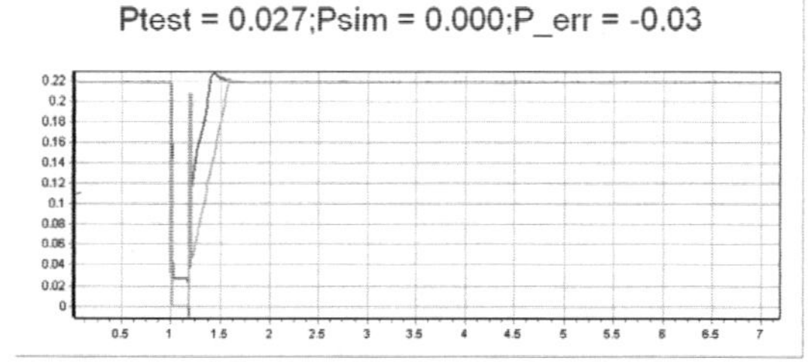


b) Active power

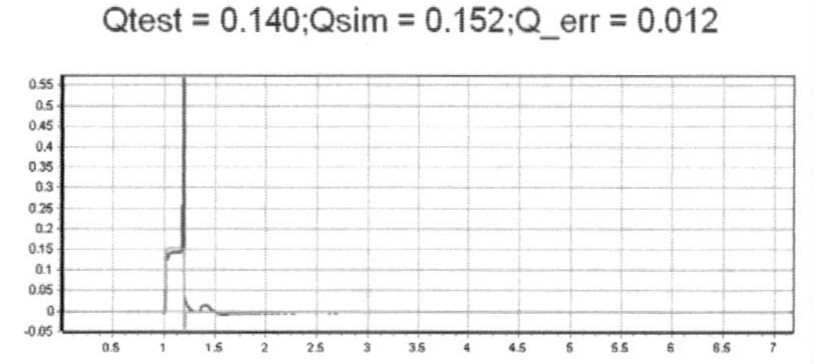


c) Reactive power

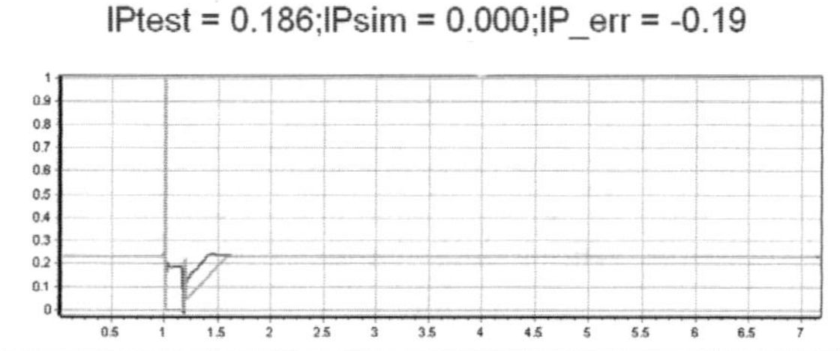


d) Active current

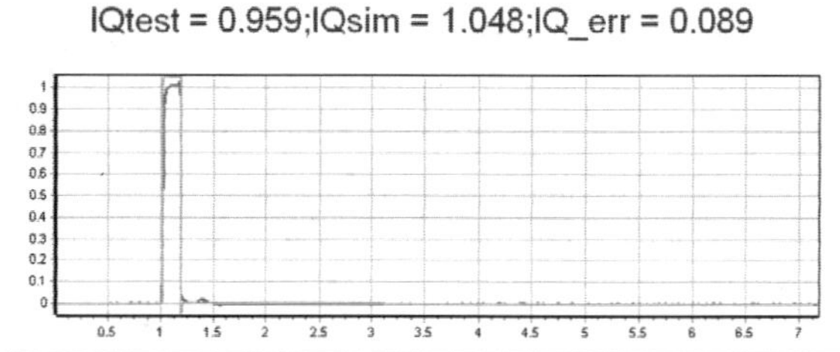


e) Reactive current

**Figure 1.** Comparison of simulation and measured data when symmetric wind voltage drops to 20%.

According to NB/T 31053-2021 "Wind Turbine Electrical Simulation Model Verification Regulations", the maximum average deviation of voltage in the fault steady-state stage is 0.05, the maximum average deviation of active power, reactive power, and reactive current is 0.15, and the maximum allowable average deviation of active current is 0.3.

The comparison results show that the proposed algorithm has better parameter identification performance than the LWPSO algorithm, and it can effectively improve the parameter identification inaccuracy problem caused by the traditional PSO algorithm, which easily falls into the local optimum. At the same time, it shows that the identification simulation results of the proposed algorithm can accurately reflect the actual operation characteristics of wind turbines, which has good practical engineering application value.

## 5. Conclusion

In this paper, a DFIG parameter identification scheme based on an improved PSO algorithm is proposed. The proposed algorithm is based on the basic particle swarm optimization algorithm. According to the optimization characteristics of two different learning factors, the Levy flight strategy is used to replace the traditional linear curve to obtain a better particle position. The simulation results show that the improved PSO algorithm can quickly identify the DFIG parameters with high accuracy, and the error of the identification parameters is small, which provides a solution for the parameter identification of LVRT.

## References

[1] Kaipeng R, Xu T, Peng W, et al. Bridging energy and metal sustainability: Insights from China's wind power development up to 2050 [J]. Energy, 2021, 227.
[2] Hamed A E S, C. H R, Elisabeth I H. Who benefits from Renewable Electricity? The differential effect of wind power development on rural counties in the United States [J]. Energy Research & Social Science, 2022, 85.
[3] Willem J Z, Andrew A. Forecasting the probability of commercial wind power development in lagging countries [J]. Cleaner Production Letters, 2022, 2.
[4] Li Li, Zheng Tiayue, Huang Shilou et al. Parameter identification method of DFIG converter control system Based on extended Kalman Filter [J]. Power Grid and Clean Energy, 2022, 38 (12): 50-60.
[5] Wang Xiaolan, Wu Dongji. Observation and Parameter Identification of Adaptive Stator Flux of Doubly-fed Generator [J]. Electric Power Transmission, 2008, (07): 59-62. DOI:10.19457/ J.1001-2095.2008.07.015.
[6] Xu Hengshan, Li Wenhao, Zhao Mingyang, et al. Characteristics Identification of Direct-driven Fan LVRT Based on Least Squares and Adaptive Snake Optimization Algorithm [J/OL]. Journal of Electric Power Systems and Automation, 2008, (07):59-62. 1-12 [2024-02-17]. https://doi.org/10.19635/j.cnki.csu-epsa.001272.
[7] Li Dan, Qin Shiyao, Li Shaolin, et al. Measurement Modeling and Transient Parameter Identification of Doubly-fed Wind Turbine LVRT Based on Chaotic Particle Swarm Optimization [J/OL]. Power China, 1-10[2024-02-17].
[8] Xue Fei, Li Hongqiang, Li Xutao et al. Control Parameter Identification Method of Doubly-fed Fan Based on LSTM Neural Network [J]. China Electric Power, 2023, 56 (06): 31-39.
[9] Zhang Zhu, Jiang Jin beauty, Xiaoping Zhang. Wolves' optimization algorithm of parameter identification of more permanent magnet synchronous motor [J]. Journal of motor and control, 2022, 26 (10): 119-129. The DOI: 10.15938 / j. MC 2022.10.014.
[10] Xie Guomin, Zhao Dejian. Multi-parameter Identification of Permanent Magnet Synchronous Motor Based on Improved Ant Lion Optimization Algorithm [J]. Journal of Electric Power Systems and Automation, 2023, 35 (06): 66-72. DOI: 10.19635/j.cnki.csu-epsa.001062.
[11] CAI G R, CHEN S L, LI S Z. Study on the nonlinear strategy of inertia weight in particle swarm optimization[C]//2008 Fourth International Conference on Natural Computation. Jinan:IEEE,2008: 683-687.

# Study on environmental monitoring classification system based on improved bayesian algorithm

Xiaoyan Guo*

Zhangye Ecological Environment Monitoring Center, Gansu Province, Zhangye, 734000, China

* Corresponding author's e-mail: 18809360829@163.com

**Abstract**. The commonly used methods for classifying water quality data are based on water quality parameters. The naive Bayesian classification method can be applied to substitute different evaluation criteria and selected water quality parameters for different water bodies. Compared to other water quality data classification methods, naive Bayesian classification methods have advantages such as simple calculation, high classification accuracy, and strong universality. However, this method overlooks the correlation between various water quality parameters and categories. To address the issues of poor universality, computational complexity, and low accuracy of traditional water quality data classification methods, a water quality data classification method based on weighted naive Bayes is proposed. This method comprehensively considers the impact of water quality attributes and their values on the classification results and replaces the original naive Bayes with weighted attribute conditional probabilities to make the classification results as close as possible to the actual category of the sample. The results indicate that this method exceeds 94% accuracy and can be directly utilized as a water quality classification module in water quality monitoring systems.

## 1. Introduction

It is imperative to actively conduct monitoring and prevention of the water environment. The classification of water quality data serves as a crucial foundation for water environment management and protection, and the quality of water quality data classification models plays a pivotal role in making decisions for water environment prevention and control [1]. In the past, the classification of water quality data relied on manual periodic sampling of water bodies, which were then taken back to the laboratory for analysis, and classification decisions were made thereafter. While this method can effectively classify according to specified standards, it consumes excessive manpower and material resources for a large number of monitoring situations and does not guarantee real-time performance [2]. With the advancement of Internet of Things technology, automatic water quality monitoring systems will become prevalent. Thus, the reasonable and accurate classification of the water quality data collected by these systems is a topic that necessitates scholarly investigation.

In recent years, some scholars have utilized Bayesian theory, based on probability and mathematical statistics, for water quality evaluation, effectively addressing the challenges encountered in water quality assessment [3-5]. In comparison with the BP neural network method, the Bayesian classification method does not necessitate a large number of water quality data samples, is straightforward to calculate, and its classification effectiveness is not inferior to the latter. The Naive Bayes classification method is one of the most widely employed classification methods, but it introduces the assumption of conditional independence [6-8]. Although, in theory, this method has the lowest misclassification rate compared to

Journal of Physics: Conference Series **2813** (2024) 012011 doi:10.1088/1742-6596/2813/1/012011

other classification methods if the condition is met, it is challenging to satisfy this condition in practical classification problems [9-10]. To mitigate the impact of the assumption of conditional independence, some scholars have commenced exploring the correlation between various attributes and employing alternative methods to enhance Naive Bayes.

To address the aforementioned limitations, this paper introduces a novel instance weighting method for the water quality classification scenario under investigation. This method assigns varying weights to each instance based on the contribution of attribute values to the classification outcomes. Additionally, in combination with the attribute weighting method, the weighted Naive Bayes final weight is determined by considering factors that influence the classification results from both the attributes themselves and their corresponding values. Subsequently, this article utilizes the data provided by the National Surface Water Quality Automatic Monitoring Station as a sample and employs the enhanced Naive Bayes classification method for the classification of water quality data. Our research is of great significance for improving water quality monitoring systems, achieving accuracy and real-time monitoring of water quality, and providing effective technical support for solving water pollution problems.

## 2. Improved naive bayesian algorithm

When calculating the conditional probability $P(a_j \mid c_i)$ using the weighted naive Bayesian classification method, different weights are assigned to each item, and its model is:

$$C_{map} = \arg\max_{c_i \in C} P(c_i) \prod_{j=1}^{N} P(a_j \mid c_i)^{w_j} \tag{1}$$

where $w_j$ indicates the weight of the attribute $A_j$, and N is the number of the condition attribute (N=4). When the weight of the condition attribute equals 1, the improved naïve Bayesian is the same as the original naïve Bayesian method.

Water quality is classified into five classes, $c_i$. The posterior probability $P(c_i \mid X)$ of dividing instance X into each category ci after introducing the weight wj can be expressed as:

$$P(c_i \mid X) = \alpha P(c_i) \prod_{j=1}^{N} P(a_j \mid c_i)^{w_j} \tag{2}$$

The weighted naive Bayesian classification method finds the term with the maximum posterior probability by calculating Equation (2), and then judges instance X as that category. The ratio of the posterior probability of instance X belonging to category ci to the sum of posterior probabilities of all categories is defined as the normalized posterior probability, which is expressed as follows:

$$P'(c_i \mid X) = \frac{P(c_i \mid X)}{\sum_{k=1}^{5} P(c_k \mid X)} \tag{3}$$

It can be seen that the larger the posterior probability $P(c_i \mid X)$ is, the greater its normalized posterior probability is. Now we select category c1 and calculate its normalized posterior probability:

$$P'(c_1 \mid X) = \frac{\alpha P(c_1) \prod_{j=1}^{N} P(a_j \mid c_1)^{w_j}}{\sum_{k=1}^{5} \left(\alpha P(c_k) \prod_{j=1}^{N} P(a_j \mid c_k)^{w_j}\right)} \tag{4}$$

 doi:10.1088/1742-6596/2813/1/012011

When considering $\frac{P(c_2)}{P(c_1)}\prod_{j=1}^{N}(\frac{P(a_j|c_2)}{P(a_j|c_1)})^{w_j}$, $\frac{P(c_2)}{P(c_1)}$ is determined by the prior probability of each class and is constant. For attribute Aj, attribute aj is determined by training sample under the prior probability $P(a_j|c_i)$ of each class. Assuming $f_i=\frac{P(a_j|c_2)}{P(a_j|c_1)}$, $f_j$ is determined. When $f_j<1$, the larger wj is, the smaller denominator is and the larger $P'(c_1|X)$ is; when $f_j>1$, the larger denominator is and the smaller the $P'(c_1|X)$ is; when $f_j=1$ $P'(c_1|X)$ is not affected by weight $w_j$, the larger the weight wj is, the greater the impact on the normalized posterior probability of instance X belonging to category ci is. In other words, attributes with larger weights contribute more to classification decisions.

Generally, the Naive Bayesian classification method based on classification probability weighting performs Naive Bayesian classification on each conditional attribute once, obtaining the probability $P_j$ of correct classification for each attribute $A_j$. Then, $P_j$ is normalized as the weight of attribute $A_j$. Considering that this weight is small, it needs to be multiplied by the number of attributes N. This article applies this method to water quality data classification, and the weight of each attribute $A_j$ is defined as:

$$w_{1,j}=\frac{NP_j}{\sum_{k=1}^{N}P_k} \tag{5}$$

Traditional water quality evaluation methods frequently assign higher weight coefficients to factors indicating higher pollution [11]. This article builds on this concept and quantifies them as part of the instance weight. In data discretization processing, the larger the attribute value is, the worse the rating of a single indicator is. The contribution of attribute $A_j$ to classification decision-making for instance X={a1, a2, a3, a4} is defined as:

$$w_{2,j}=\frac{a_j}{\frac{1}{N}\sum_{k=1}^{N}a_k} \tag{6}$$

Considering the impact of attributes and their values on the classification results, we quantify them as the final weight wj of instance attributes. Therefore, we defined $w_j$ as:

$$w_j=\frac{w_{1,j}+w_{2,j}}{2} \tag{7}$$

In machine learning algorithms, performance evaluation for a binary classification problem typically uses metrics of accuracy, precision, recall, and F1 value. In the Binary confusion matrix, TP indicates the condition that both the prediction and observed value are true, prediction is negative and observed is positive for FN, prediction is positive and observed is negative for FP, and prediction is negative and observed is negative for TN. Finally, the sum of TP, FP, FN, and TN is the total sample.

Accuracy represents the ratio of the number of samples successfully predicted as positive by the classifier on the test set to the total number of samples predicted as positive, defined as:

$$P=\frac{TP}{TP+FP} \tag{8}$$

The five-fold cross-validation process involves dividing the existing water quality dataset into five groups on average, ensuring the consistency of the category distribution in each group, and alternating between using the combined data of four groups as the training set and the remaining one group as the testing set. The classification performance of each method on water quality data is then evaluated based on the results of five rounds of testing.

To assess the effectiveness of the method proposed in this paper for classifying water quality data, four sets of experiments were designed for comparative analysis. These experiments involved using the standard Naive Bayes classification method (SNBC), the Naive Bayes classification method based on classification probability weighting (CNBC), the Attribute Value Frequency-Based Instance Weighting Filter for Naive Bayes (FNBC), and the method proposed in this paper for water quality data classification.

The data in this study is sourced from real-time monitoring data from the National Surface Water Quality Automatic Monitoring Station (National Water Control Station), which publishes 9 indicators including water temperature, pH, and dissolved oxygen, according to the National Surface Water Quality Monthly Report in May 2022, among the 3573 national examination sections monitored. Considering the limited amount of data on inferior Class V water quality, this article divides it into Class V datasets in the experiment.

## 3. Experimental results

This study uses a five-fold cross-validation method to classify water quality data samples using four methods, and the classification results are statistically analyzed using a confusion matrix table. According to Equations (3.15) and (3.16), the accuracy rates of the four methods under each category are shown in Figure 1, and the recall rates of the four methods under each category are shown in Figure 1.

The accuracy of the four methods for classifying each water quality category has almost reached over 90%. Among them, the accuracy of the method proposed in this article is higher than that of the standard naive Bayesian classification method for each category. Only in categories 5 and 3, the accuracy is lower than that of the CNBC and FNBC methods, respectively. In addition, the recall rate of the method proposed in this article is high/equal to the other three naive Bayesian classification methods in each category (Figure 2).

The overall performance indicators of each method for water quality sample classification and the comparison of performance indicators of the four methods in water quality classification are shown in Table 1.

**Table 1.** Performance indicators of four methods in water quality classification (%). Standard Naive Bayes classification method (SNBC), the Naive Bayes classification method based on classification probability weighting (CNBC), the Attribute Value Frequency Based on Instance Weighting Filter for Naive Bayes (FNBC), and the method proposed in this paper.

| **Method** | Accuracy | Precision rate | Recall rate | F1 |
|---|---|---|---|---|
| **SNBC** | 94.5 | 94.3 | 91.5 | 93.2 |
| **CNBC** | 94.6 | 94.8 | 91.6 | 93.1 |
| **FNBC** | 95.6 | 94.5 | 92.6 | 93.9 |
| **Our method** | 97.1 | 96.6 | 94.6 | 95.8 |

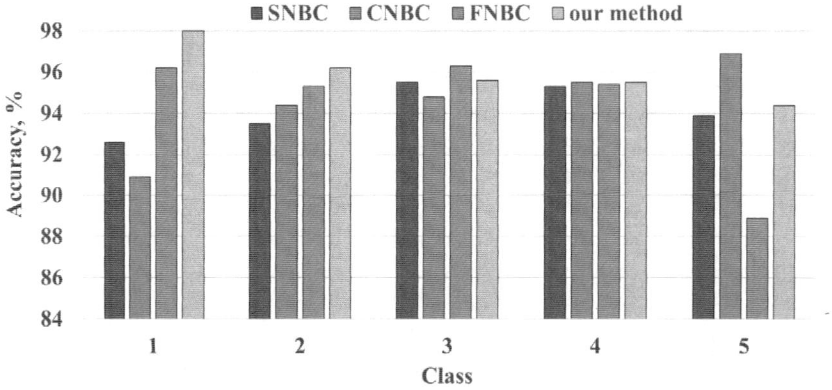


**Figure 1.** The accuracy of four methods under each category. Standard Naive Bayes classification method (SNBC), the Naive Bayes classification method based on classification probability weighting (CNBC), the Attribute Value Frequency Based on Instance Weighting Filter for Naive Bayes (FNBC), and the method proposed in this paper.

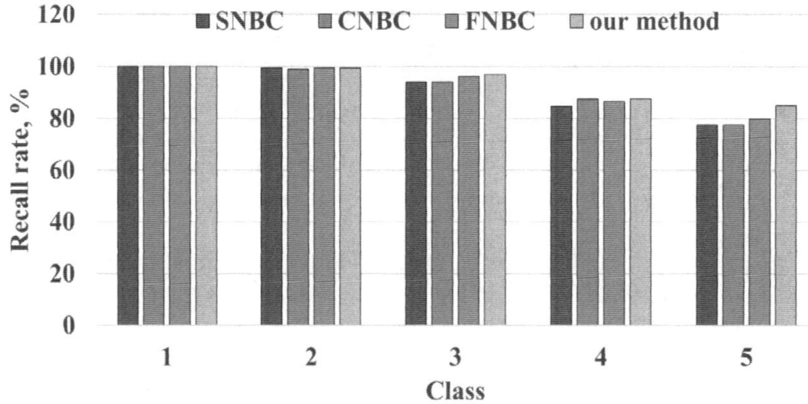


**Figure 2.** Recall rates of four methods under each category. Standard Naive Bayes classification method (SNBC), the Naive Bayes classification method based on classification probability weighting (CNBC), the Attribute Value Frequency Based on Instance Weighting Filter for Naive Bayes (FNBC), and the method proposed in this paper.

The proposed weighted Naive Bayes classification method outlined in this paper achieves an accuracy, precision, recall, and F1 value of 96.0%, 95.9%, 93.8%, and 94.8% for the classification of water quality samples, respectively. All performance indicators exceed 90%, indicating strong classification results. When compared with the standard Naive Bayes classification method, CNBC, and FNBC methods, the approach presented in this paper enhances the accuracy of water quality sample classification by 1.8%, 1.6%, and 0.8%, precision by 2%, 1.6%, and 1.8%, recall by 3.1%, 2.4%, and 1.5%, and F1 values by 2.3%, 1.9%, and 1.5%, respectively. The comparative analysis demonstrates that the method proposed in this paper significantly enhances the performance indicators for water quality sample classification when compared to the other three methods, rendering it more suitable for water quality classification.

For each sample instance, the Naive Bayes classification method computes the posterior probability of each attribute within each category and selects the term with the highest posterior probability as the classification result for that instance. Consequently, as the Naive Bayes classification method calculates the posterior probability of each test sample in the actual category, a higher value corresponds to increased credibility in the classification results. To emphasize the influence of weighted Naive Bayes classification methods on the posterior probability of actual categories, we will tabulate the number of errors in standard Naive Bayes classification, randomly select 10 instances, and gather the posterior probabilities of these samples under each category using the four methods. To facilitate analysis, the normalized posterior probability under the actual category was calculated. The comparative analysis of the normalized posterior probabilities of the actual category is presented in Figure 3.

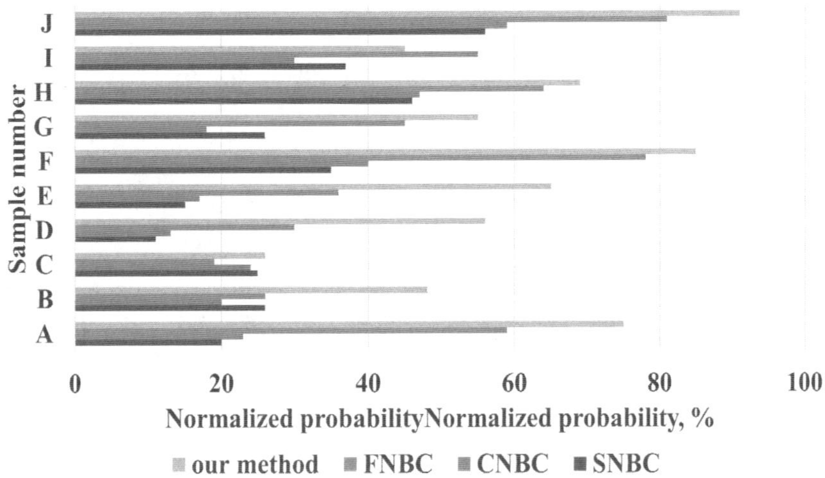


**Figure 3.** Comparison of normalized posterior probabilities of actual categories. Standard Naive Bayes classification method (SNBC), the Naive Bayes classification method based on classification probability weighting (CNBC), the Attribute Value Frequency Based on Instance Weighting Filter for Naive Bayes (FNBC), and the method proposed in this paper.

The traditional water quality data classification method has problems such as poor universality, cumbersome calculation, and low accuracy. This paper proposes a water quality data classification method based on weighted naive Bayes. With existing samples, this method first calculates the correlation between each water quality attribute and category, then considers the impact of each attribute value on the classification results. Finally, the weight values of each instance attribute are defined by combining two factors, and the weighted attribute conditional probability is used to replace the original naive Bayes, making the classification results as close as possible to the actual category of the sample. Through simulation comparison with other naive Bayesian classification methods, the results show that this method has better classification performance for water quality data and can be directly used in the water quality classification module of water quality monitoring systems.

## 4. Conclusion

This article explores a classification method for water quality data based on Naive Bayes. However, the conditional independence assumption of Naive Bayes significantly impacts the classification performance of water quality data. Thus, an instance weighting method is proposed to address this limitation. The findings demonstrate that the proposed method offers enhanced classification performance for water quality data, achieving accuracy, precision, recall, and F1 values of 96.5%, 96.2%, 94.1%, and 95.2%, respectively. This method can be directly employed in the water quality classification module of the water quality monitoring system in the future. The research results of this article are of great significance for a comprehensive water quality monitoring system to achieve accuracy and real-time monitoring of water quality, and provide effective technical support for solving water pollution problems. At the same time, it is important to establish a lightweight water quality monitoring node based on NB IoT.

**References**

[1] Liu W, Wang S, Zhang L, et al.(2015). Water Pollution Characteristics of Dianchi Lake and the Course of Protection and Pollution Management. Environmental Earth Sciences, 74:3767-3780. DOI:10.1007/s12665-015-4152-x.
[2] Jakositz S, Pillsbury L, Greenwood S, et al.(2019). Protection Through Participation: Crowdsourced Tap Water Quality Monitoring for Enhanced Public Health. Water Research, 169:24-43.DOI:10.1016/j.watres.2019.115209.
[3] Boni A, Bianchi V, Ricci A, et al.(2021). NB-IoT and Wi-Fi Technologies: An Integrated Approach to Enhance Portability of Smart Sensors. IEEE Access, 9:74589-74599. DOI:10.1109/ACCESS.2021.3082006.
[4] Sui H, Zheng G, Zhou J, et al. Application of NB-IoT Technology in City Open Water Monitoring.2020 6th International Symposium on System and Software Reliability (ISSSR), Chengdu, China, vol.8, pp. 95-98, July 2020.
[5] Batista L V. Turbidity Classification of the Paraopeba River Using Machine Learning and Sentinel-2 Images. IEEE Latin America Transactions, vol. 20, pp. 799-805, April 2022.
[6] Yu L, Gan S, Chen Y, et al. Correlation-Based Weight Adjusted Naive Bayes. IEEE Access, vol. 8, pp.51377-51387, July 2020.
[7] Jamieson R, Gordon R, Joy D, et al. Assessing Microbial Pollution of Rural Surface Waters.Agricultural Water Management, vol.70, pp.1-17, April 2004.
[8] Siva Subramanian R, Prabha D. Customer Behavior Analysis Using Naive Bayes with Bagging Homogeneous Feature Selection Approach. Journal of Ambient Intelligence and Humanized Computing, vol. 12, 5105-5116, July 2021.
[9] Chen Xi, Zhang Kun A classifier learning method based on tree enhanced naive Bayes. Journal of Electronics and Information Technology, vol. 41, pp.2001-2008, June 2019.
[10] Xu W, Jiang L, Yu L. An Attribute Value Frequency-Based Instance Weighting Filter for Naïve Bayes. Journal of Experimental & Theoretical Artificial Intelligence, vol. 31, pp. 225-236, May 2019.
[11] Hou W, Sun S, Wang M, et al. Assessing Water Quality of Five Typical Reservoirs in Lower Reaches of Yellow River, China: Using A Water Quality Index Method. Ecological Indicators, vol.61, pp.309-316,2016.

Journal of Physics: Conference Series **2813** (2024) 012012 doi:10.1088/1742-6596/2813/1/012012

# Differential privacy protection algorithm for large data sources based on normalized information entropy Bayesian network

**Guangyuan Ni* and Jiaxin Sun**

China Ship Scientific Research Center, WuXi, 214000, China

*Corresponding author's e-mail: nigy@cssrc.com.cn

**Abstract:** With the rapid development of the Internet, privacy protection in big data has become an important research hotspot in the field of information security. Among them, differential privacy protection is the most widely used big data privacy protection technology today. However, the existing differential privacy protection technology cannot fully and effectively deal with the problem of high-dimensional privacy big data. Although the private data publishing method based on a Bayesian network can effectively deal with the conversion problem of high-dimensional data sets to low-dimensional data sets, this method also has some defects and deficiencies. Therefore, this paper proposes a large data source differential privacy protection algorithm (NIE-PrivBayes algorithm) based on normalized information entropy Bayesian network and proposes three improvements to the original PrivBayes algorithm. First, the original data set is protected by differential privacy on the client side. Secondly, the Bayesian network is constructed by the joint probability distribution obtained by the EM algorithm. Finally, the existing Bayesian network construction algorithm is optimized in the aspects of normalized information entropy. Based on significantly reducing the computational overhead, the effectiveness and availability of publishing high-dimensional data sets are realized under the premise of differential privacy protection.

## 1. Introduction

At present, high-dimensional big data is the most critical part of an information network. Many data research organizations need to release large amounts of raw data for research analysis or other applications. However, high-dimensional privacy big data often contains sensitive personal privacy information, so it needs to be processed by special protection technology. At present, these privacy protection processing technologies are roughly divided into three categories: 1) Data distortion-based (random disturbance [1-3], data injection noise [4-6], etc.); 2) Data encryption (DES encryption, RSA encryption[7-8], etc.); 3) Restricted conditions (K-Anonymous Privacy Protection Act [9]). Differential privacy protection is the most widely used data privacy protection technology by adding noise to the original data. However, this technology can not deal with the issue of publishing high-dimensional privacy data comprehensively and effectively. Zhang et al. proposed the PrivBayes data publishing method [10], which realizes the transformation of high-dimensional data sets to low-dimensional data sets based on Bayesian networks [11] and adopts differential privacy protection processing technology in low-dimensional data sets. However, there are still defects and deficiencies.

In this paper, we propose a differential privacy protection algorithm for large data sources (NIE PrivBayes algorithm) based on normalized information entropy Bayesian network. The NIE-PrivBayes algorithm is mainly divided into four steps: First, each attribute value is converted into a random binary string, and then sent to the central server to achieve differential privacy protection at the data source. Secondly, the EM algorithm is used to calculate the joint probability distribution of multi-dimensional attributes to the maximum extent. Then, the first attribute node is determined by normalized information entropy, and the mutual information calculation based on statistics is implemented based on the EM algorithm, building a higher-quality Bayesian network. Finally, the data set is resynthesized by combining the probability distribution for data publication. Finally, based on significantly reducing the computing overhead, the effective balance between privacy and availability of high-dimensional big data publishing is achieved. Figure 1 shows the basic framework and ideas of this paper.

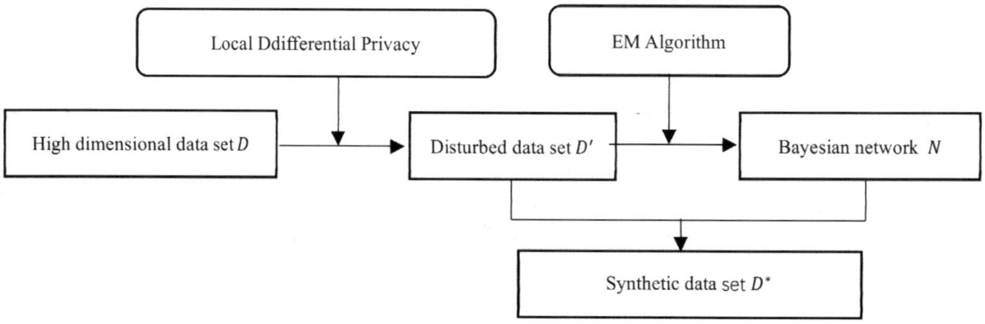


**Figure 1.** Flow chart of high-dimensional data source differential privacy protection.

## 2. Related work

*2.1 Problem description*

Let's assume that there are n users, each with d attributes. The user's raw data set is $U = \{u^1, u^2, \dots, u^n\}$, where $u^i$ represents the data record of the $i$ user, whose attribute values are expressed as $u^i = \{u_1^i, u_2^i, \dots, u_d^i\}$. On the other hand, the set of attributes of the data set $U$ is represented by $A = \{a^1, a^2, \dots, a^d\}$. The value of each attribute is expressed as $a^i = \{a_k^i, a_k^i, \dots, a_k^i\}$. The attribute range is $\Omega = \{\Omega_1, \Omega_2, \dots, \Omega_d\}$. Our ultimate goal is to have the resultant data set $X^*$eventually published on the server have a similar distribution to the given original data set $X$. That is, the joint probability distribution of the synthetic data set $X^*$ on the attribute set A satisfies: $P_{X^*}(a^1, a^2, \dots, a^d) = P_X(a^1, a^2, \dots, a^d)$.

*2.2 Data source differential privacy protection*

Firstly, the data source differential privacy protection is implemented for the original high-dimensional data set. The goal is to prevent the data set from being attacked by an attacker during transmission to or on the server. The data source differential privacy protection algorithm is shown as follows.

**Step 1:** Binary processing. Suppose that the original data of the $i$ th user $u^i$ is recorded as $u^i = \{u_1^i, u_2^i, \dots, u_d^i\}$, the $i$ th original attribute is recorded as $a^i$, and each attribute takes the value $a^i = \{a_k^i, a_k^i, \dots, a_k^i\}$, and for each attribute $a^i$, the range size is $|\Omega_i|$, its position is set to 1 and the rest to 0, and finally the binary string $\hat{s}_i$ of length $|\Omega_i|$ is formed.

**Step 2:** Bit flip. After binary processing of the original data, the binary string $s_j^i$ is obtained, and then each bit of it is randomly answered with the probability of $f$, and the real answer with the probability of $1 - f$. $f$ is a user-adjustable parameter that controls the level of privacy protection.

**Step 3:** Binary string join. By connecting the $j$ attribute binary strings $s_j^i$ of $i$ users with the binary string $s_j^i$ obtained after random flip, a binary string $\hat{s}_i$ with the length $\sum_{j=1}^{d} |\Omega_j|$ is obtained and sent to the server to construct the Bayesian network. In this case, the last returned binary string $\hat{s}_i$ has differential privacy protection on the local client.

### *2.3 EM algorithm*

In this paper, the EM algorithm is used to calculate the joint probability distribution of $d$-dimensional attributes.

Let $D$ be represented as the index set of the attribute set $A$, i.e. $\mathrm{D} = \{1,2,\ldots,d\}$, the value of the attribute $a^i$ is denoted as $\omega_i$. Use $\mathrm{P}(\omega_1\omega_2 \ldots \omega_d)$ or $\mathrm{P}(\omega_\mathrm{D})$ to represent the joint probability distribution of $d$-dimensional attributes.

**Step 1:** Parameter initialization. Let the initial joint probability be $P_0(\omega_\mathrm{D})=\frac{1}{\prod_{j=1}^{d}|\Omega_j|}$.

**Step 2:** Calculation of the conditional probability. The conditional probability denoted $P(\hat{s}_1^i\hat{s}_2^i \ldots \hat{s}_\mathrm{d}^i|\omega_1\omega_2 \ldots \omega_d)$. According to the bit flipping algorithm, finally, it is converted to the binary string $s_j^i$. It follows that:$P(\hat{s}_1^i\hat{s}_2^i \ldots \hat{s}_\mathrm{d}^i|\omega_1\omega_2 \ldots \omega_d) = \prod_{b=1}^{|\Omega|}(\frac{f}{2})^{s_\mathrm{D}[b]}(1-\frac{f}{2})^{1-s_\mathrm{D}[b]}$.

**Step 3:** Expectation maximization estimation. EM each iteration is divided into a desired step (E step) and a maximum step (M step).

(1) E Step: The posterior probability is calculated. From the Bayes probability get:$P_k(\omega_1\omega_2 \ldots \omega_d|\hat{s}_1^i\hat{s}_2^i \ldots \hat{s}_\mathrm{d}^i) = \frac{P_{k-1}(\omega_1\omega_2\ldots\omega_d)P(\hat{s}_1^i\hat{s}_2^i\ldots\hat{s}_\mathrm{d}^i|\omega_1\omega_2\ldots\omega_d)}{\sum_{\omega_1}\cdots\sum_{\omega_1}P_{k-1}(\omega_1\omega_2\ldots\omega_d)P(\hat{s}_1^i\hat{s}_2^i\ldots\hat{s}_\mathrm{d}^i|\omega_1\omega_2\ldots\omega_d)}$

(2) M Step: iteratively update parameter $P_k(\omega_\mathrm{D})$. The joint probability of a new round of $d$-dimensional attributes replaces the prior probability of the previous round with $P_k(\omega_\mathrm{D})$, and then returns the E step.

Steps E and M iterated until $max_{\omega_\mathfrak{D}}P_k(\omega_\mathfrak{D}) - max_{\omega_\mathfrak{D}}P_{k-1}(\omega_\mathfrak{D}) \leq \delta$.

Through the above three steps, the joint probability of $d$-dimensional attributes can be obtained. Bayesian networks can then be constructed.

### *2.4 Bayesian network improvement*

In this paper, some improvements are made to the shortcomings of the traditional Bayesian algorithm. Improvement point 1 is to select initial attribute nodes through normalization of information entropy, which can improve the probability distribution of initial attribute nodes, so the finally constructed Bayesian network can restore the correlation of the original attribute set to the greatest extent. Improvement point 2 is further optimization on the mutual information function, including two optimization points: one is to add the information comparison Table 1, which can reduce repeated calculation and thus reduce the calculation cost; the other is to realize the statistical calculation mutual information of the probability estimation model based on EM. This approach significantly reduces computational overhead and memory consumption. Through the above two improvements, higher-quality Bayesian networks can be constructed quickly and efficiently.

**Table 1.** Improved Bayesian network algorithm NIE-PrivBayes.

| Algorithm 1: Improved Bayesian network algorithm NIE-PrivBayes |
|---|
| Input: Data set $D$, Original attribute set $\mathrm{A} = \frac{\{a^1,a^2,\ldots,a^d\}}{}$. The maximum penetration $k$. |
| Output: Bayesian network $N$. |
| 1: From attribute set A, the normalized information entropy of each attribute is calculated, the attribute $a^1$ with the highest normalized information entropy is selected as the initial node of $N$, $a^1$ is added to $\Gamma$, and the AP pair $(a^1,\emptyset)$ is added to $N$; |
| 2: For $i = 2$ to $d$ do |
| 3: Initialize $\Pi = \emptyset$; |
| 4: If $\|\Gamma\| > \mathrm{k}$ |
| 5: For $\forall\mathrm{A} \in \mathrm{A}/\Gamma$ and $\forall\{\Pi \in \mathrm{C}_\Gamma^k\|A1 \in \Pi\}$, $(A,\Pi)$ and $(\mathrm{A1},\Pi1)$ are added to the $\Gamma$, based on improved mutual information function to calculate the mutual information$\mathrm{I}(A,\Pi)$; |
| 6: Else |

7: For $\forall A \in A/\Gamma$ and $\Pi \in C_{\Gamma}^{k}$, $(A, \Pi)$ is added to the $\Gamma$, based on improved mutual information function to calculate the mutual information$I(A, \Pi)$;

8: End if

9: The child-parent pair is selected with the maximum mutual information function value I, and it is labeled as $(A, \Pi)$

10: The child-parent pair $(A1, \Pi1)$ is selected with the maximum mutual information function value I1, where $A1$ is not equal to $A$;

11: $A$ is added to$\Gamma$and $(A, \Pi)$ is added to network $N$

*2.5 Synthetic dataset*

Given $A$ Bayesian network $N$, the joint probability $P(A)$ of all attributes is represented as follows:$P(A) = P(a^1, a^2, \dots, a^d) = P(a^d | a^1, \dots, a^{d-1}) \dots Pr(a^2|a^1)P(a^1) = \prod_{i=1}^{d} P(a^i|\Pi_i)$.

The data on each attribute $a^i$ can be independently sampled according to $P(a^i|\Pi_i)$, and then the synthetic attribute set can be generated. The specific process is as follows:

(1) The initialized node data are set and the attribute node data are sampled, denoted as M.

(2) The other attribute nodes are sampled. The conditional probability distribution $P(a^i|\Pi_i)$ is calculated.

(3) Iterating through the loop (2) until all attribute nodes are sampled. Finally, the sampled data from all nodes are synthesized into a new $M \times d$ dataset.

## 3. Experimental evaluation

*3.1 Experimental data set and parameter setting*

The Audlt dataset is used, mainly contains 45,222 personal information records from the 1994 U.S. Census, with 15 attributes including salary, college, etc.

In the simulation process of the NIE-PrivBayes algorithm, the effect of the algorithm is compared and analyzed mainly through data privacy protection and the Bayesian network.

*3.2 Bayesian network analysis (Isum)*

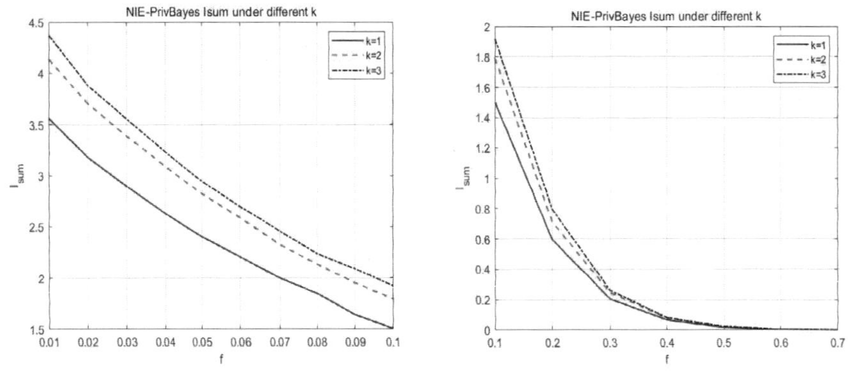


**Figure 2.** NIE-PrivBayes mutual information entropy sum under different k.

As can be seen from Figure 2, with the increase of bit flipping probability f value, Bayesian network mutual information and Isum value gradually decrease, and network attributes tend to be independent, conforming to the theoretical analysis situation.

In addition, with the specified f, the Isum value of the network increases with the increase of the maximum number of parent nodes k, indicating that with the increase of the number of parent nodes of child nodes, the correlation expression between attributes is more accurate, and the quality of the network construction is also higher.

*3.3 Data privacy analytics (epsilon)*

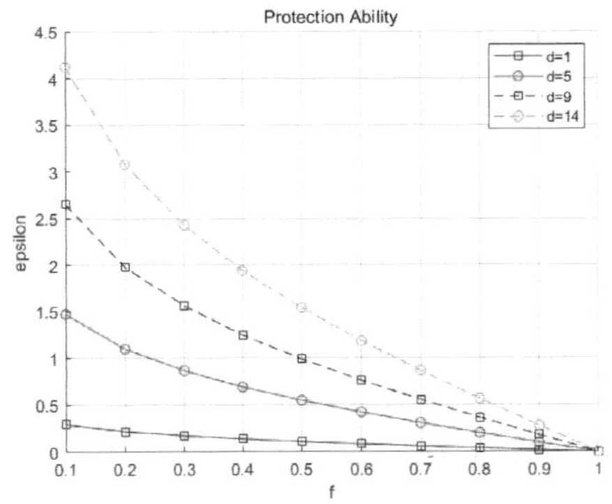


**Figure 3.** The protection ability in different dimensions d.

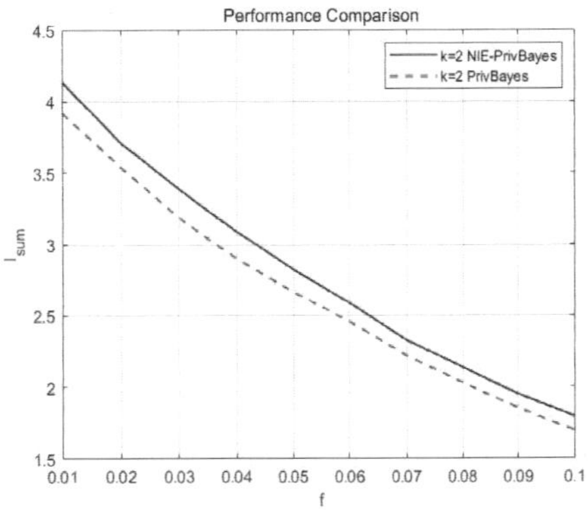


**Figure 4.** Performance comparison between NIE-PrivBayes and PrivBayes.

As can be seen from Figure 3, for a specific dimension, with the increase of f, the epsilon value gradually decreases and data privacy performance is better. Under the comparison effect of different dimensions, when the f increases, the epsilon value also increases with the increase of data dimension, which is in line with reality.

From Figure 4, under the premise that the specified degree value k=2, after the normalized information entropy selection of the first attribute node and the optimization of the mutual information function, the Isum of NIE-PrivBayes is larger, indicating that the attribute correlation of the Bayes network constructed by the NIE-PrivBayes algorithm is higher.

## 4. Conclusion

This paper mainly introduces the differential privacy protection algorithm for large data sources (NIE-PrivBayes algorithm) based on normalized information entropy Bayesian network. It mainly introduces the shortcomings of the original PrivBayes algorithm and puts forward three improvements: Firstly, the differential privacy protection of data sources is introduced. Secondly, the EM algorithm is introduced. Finally, the existing Bayesian network construction algorithm is optimized in three parts: selecting the first attribute node based on the normalized information entropy, adding the information comparison table, and realizing the mutual information value calculation based on the statistical probability of the EM algorithm. Based on the above three improvements, the NIE-PrivBayes algorithm significantly reduces the computing overhead and improves the privacy protection and availability of data.

## References

[1] Zhang L., Liu Y., Wang RC. (2016) Location Publishing Technology Based on Differential Privacy-preserving for Big Data Services[J]. Journal on Communications, 37(9): 46-54.
[2] Wang L., Wang WP., Meng Dan. (2016) Privacy Preserving Data Publishing via Weighted Bayesian Networks[J]. Journal of Computer Research and Development, 53(10): 2343-2353.
[3] Rui C., Qian X., Yu Z., Xu JL. (2015) Differentially private high-dimensional data publication via sampling-based inference. In Proceedings of SIGKDD. 129–138.
[4] Chen Y. Ashwin, Machanavajjhala. 2015. On the privacy properties of variants on the sparse vector technique. CoRR abs/1508.07306 (2015).
[5] Chen R., Xiao Q., Zhang Y., Xu J. (2015) Differentially private high-dimensional data publication via sampling-based inference. In: Proceedings of the 21st ACM SIGKDD international conference on knowledge discovery and data mining, pp 129–138
[6] Chen H., Cai M., Chen X. 2023. Privacy Protection Method for Cellular Signaling Data Based on Genetic Algorithm[J]. Journal of Transportation Engineering, Part A: Systems.

[7] Wu ZB., Xiao WH., Deng H., Yang L., 2022. Research on data privacy protection methods based on genome-wide association study[J]. International Journal of Embedded Systems.
[8] Wang ZL., Ju AY. 2022. A novel privacy protection method based on node segmentation for social networks[J]. International Journal of Communication Networks and Distributed Systems.
[9] Wang SJ., Wang LL. 2020. Particle Gibbs Sampling for Bayesian Phylogenetic inference.[J]. Bioinformatics. Oxford, England.
[10] Ju CH., Gu QY., Wu GX., Zhang SG. 2020. Local Differential Privacy Protection of High-Dimensional Perceptual Data by the Refined Bayes Network[J].
[11] Yang GM., Ye XX., Fang XJ., Wu RS., Wang L. 2020. Associated Attribute-Aware Differentially Private Data Publishing via Microaggregation[J].IEEE Access.

Journal of Physics: Conference Series **2813** (2024) 012013 doi:10.1088/1742-6596/2813/1/012013

# Blockchain-based anonymous (IOT) data integrity auditing scheme in cloud environments

**Jianhong Zhang and Jianan Zhang***

North China University of Technology, Beijing, 100144, China

*Corresponding author's e-mail: zhangjiananup0714@163.com

**Abstract.** The Internet of Things (IoT) facilitates the interaction, data collection, and exchange between devices and the network. With an increasing proliferation of IoT devices, the volume of data produced escalates at an exponential rate. Considering IoT's constrained computational and storage capabilities, the necessity for cloud data storage becomes imperative to enhance service quality. Scholars have introduced various auditing mechanisms, within which the role of a third-party auditor (TPA) emerges as a distinctive attribute in certain methodologies. Nonetheless, assuring the security of a centralized TPA poses significant challenges, a gap that blockchain technology seeks to bridge. Addressing this issue, our proposition is the Blockchain-based Anonymous Data Integrity Auditing (BADIA) for IoT within cloud environments, employing a federated blockchain approach. Our scheme has the following characteristics: conditional anonymity, fairness, accountability, and security. In conclusion, a comprehensive assessment, considering both security and performance metrics, demonstrates that the proposed scheme is robust and effectively alleviates the computational load for data proprietors who possess limited processing capabilities. This analysis ensures that our methodology not only adheres to stringent security standards but also optimizes the efficiency of data management for those with constrained computational resources.

## 1. Introduction

The Internet of Things (IoT) constitutes an extensive network of linked devices, pivotal across multiple sectors including transportation, healthcare, and agriculture[1]. As IoT data burgeons exponentially, cloud platforms emerge as essential for providing the computational and storage capacities required for IoT devices to delegate their data[1][2]. Nonetheless, safeguarding the integrity and accessibility of IoT data within cloud environments remains a focal point of scholarly inquiry.

Several data integrity auditing schemes have been proposed, but they may have trust issues, data privacy risks, and single points of failure. They often rely on complex management of public key certificates [3]. To surmount the identified obstacles, this paper presents a scheme predicated on blockchain technology for the anonymous auditing of data integrity within the Internet of Things (IoT) environments [5]. The proposed methodology integrates homomorphic hashing with identity-based cryptographic mechanisms to establish an efficient system for the verification of data integrity[6][7]. Utilizing blockchain technology, this strategy renders third-party auditors obsolete by facilitating the direct verification of the integrity of data transferred to cloud storage solutions. Tailored specifically for the constraints of IoT devices, which are often limited in computational power, this architecture provides a secure and effective means for data integrity verification [9]. In essence, the scheme is designed to tackle issues of trust, privacy, and operational efficiency in the auditing of IoT data integrity through the strategic application of blockchain technology and identity-based cryptographic techniques [10].

## 2. Preliminaries

*2.1. Hardness problems bilinear and pairing*
In the realm of cryptography, bilinear pairing, comparable to the challenge posed by the elliptic curve discrete logarithm problem, facilitates the simplification of intricate group problems into more manageable forms. It involves a map $e: G_1 \times G_1 \rightarrow G_2$ between two cyclic additive groups of the same prime order \$p\$ characterized by bilinearity, non-degeneracy, and computability.

Bilinear function: $e(g_1{}^a, g_2{}^b) = e(g_1, g_2)^{ab}$.

Non-degenerate: ensuring non-trivial results $e(g_1{}^a, g_2{}^b) \neq 1 \in \mathbb{G}_2$.

Computability: efficient calculation in polynomial time.

## 3. System and threat model

*3.1. System model*
The architectural framework of the Blockchain-based Anonymous Data Integrity Auditing (BADIA) scheme for cloud storage is illustrated. This comprehensive system encompasses four distinct entities: the Public Key Generator (PKG), the Data Owner (DO), the Cloud Service Provider (CSP), and the Blockchain network.

PKG: It generates system-wide parameters, anonymous identities, and public-private key pairs for data owners.

DO: Anonymously outsources IoT data to cloud servers, verifies data integrity, generates digital signatures and IVA, and uploads them to CSP and blockchain securely.

CSP: It stores IoT data and digital signatures, accepts integrity audit challenges from the blockchain, and provides verification proofs.

Blockchain: It stores aggregated signatures and random numbers, publicly verifies data integrity, and facilitates integrity audit tasks for DOs and authorized users.

## 4. Our proposed construction

*4.1. Our construction*
This study introduces a framework dedicated to the verification of medical data integrity within an externally managed cloud storage setting, employing blockchain-based conditional identity anonymization. The scheme includes several components, namely, Setup, SetSecretValue, SigGen, Challenge, ProofGen, and Verify. The specific program flow is as follows.

- Setup($\zeta$)$\rightarrow$(msk, params): It is provided with a security parameter $\zeta$. PKG generates system parameters. Concretely, in the cryptographic schema, the Parameter Generation Key (PKG) selects two cyclic groups, $G_1$ and $G_2$, each possessing an identical prime order denoted as $p$. PKG chooses a safe bilinear map $e: G_1 \times G_1 \rightarrow G_2$. We assume that $g$ is a generator of $G_1$, and PKG set two safe and collision-resistant universal hash functions, $H_1: \mathbb{G}_1 \times \mathbb{G}_1 \times \{0,1\}^* \rightarrow \{0,1\}^\rho$, $H_2: \mathbb{G}_1 \times \{0,1\}^\rho \rightarrow \{0,1\}^\rho$ as well as two homomorphic hash functions, $H_3: \mathbb{Z}_p \rightarrow \mathbb{G}_1$, $H_4: \mathbb{Z}_p \times \mathbb{G}_1 \times \mathbb{Z}_p \rightarrow \mathbb{G}_1$. Moreover, we randomly select $x \leftarrow Z_p^*$ as the master private key \$msk\$, calculate $mpk = g^x$ as the master public key of PKG, and randomly and uniformly select $\omega \leftarrow G_1$. The system parameters are lastly made public by PKG, while the master private key $x$ is secretly and securely stored. $params = (e, \mathbb{G}_1, \mathbb{G}_2, g, p, \omega, mpk, H_1, H_2, H_3, H_4)$.

- $SetSecretValue \rightarrow (PID, Sig_i)$ : This is an algorithm for PKG to generate a pseudonymous identity $PID$ and corresponding signed private key $Sig_i$ for the DO's real identity $ID \in \{0,1\}^{\rho}$. The PKG randomly chooses a secret value $s \leftarrow Z_p^*$, and calculates the pseudonymous identity $PID$ for the DO using the master private key msk. $PID = (PID_a, PID_b)$ , $PID_a = g^s$ , $PID_b = ID \oplus H_1(PID_a^x \| mpk \| Time)$ represents the effective period during which a DO can maintain their anonymity. Finally, PKG uses the master private key msk to compute the signing private key $SK_{PID}$ corresponding to the pseudonymous identity $SK_{PID} = (s+x)H_2(PID)$.
- $SigGen \rightarrow Sig_i$: This algorithm is utilized by DO to generate digital signatures based on IoT device data files and calculate information that supports integrity audit verification. The DO first slices the data F into n blocks using data slicing, which is denoted as $F[1], F[2], \cdots, F[i], F[i] \in Z_P^* (i = 1, 2, \cdots n)$ . The file name is $F_{id} \in Z_P^*$ . Then, the user randomly chooses these values $\sigma, \kappa, \gamma \in Z_P, t \in Z_P^*$, calculating seed information $\varphi_1 = \sigma \cdot \kappa \cdot \gamma$, $\varphi_2 = \sigma^2 \cdot \kappa^2 \cdot \gamma, \cdots \varphi_n = \sigma^n \cdot \kappa^n \cdot \gamma$ , $Y = g^t$ and generating integrity validation auxiliary information $IVA = \{\sigma, \kappa, \gamma, t, H_3(\varphi_1), H_3(\varphi_2), \cdots, H_3(\varphi_n)\}$. Furthermore, the DO computes the digital $Sig_i$ for each data block using $SK_{PID}$ , $T_i = (\omega^{F[i]} H_4(Fid \| PID \| i))^{SK_{PID}}$ , $R_i = (\varphi_i \cdot H_3(PID))^t$ , and $Sig_{i=} R_i \cdot T_i$. Finally, the DO uploads the signature and file blocks $(\{F[i]\}_{1 \le i \le n}, \{Sig_i\}_{1 \le i \le n})$ to the CSP, sends the $IVA$ to be stored on the blockchain via a secure channel, and deletes the local copy.
- $Challenge \rightarrow chal$ : This methodology is employed upon a user's issuance of a mandate, enabling the blockchain to conduct an integrity audit on data that has been externalized to a Cloud Service Provider (CSP). Subsequently, the blockchain formulates a challenge message, which is then dispatched to the CSP. The blockchain randomly selects a subset $J = \{j_1, j_2, \cdots j_m\}$ containing $m$ elements from the set $\{1, 2, \cdots, n\}$ and assigns a random coefficient $v_j \leftarrow Z_p$ . It then sends the resulting challenge information $chal = \{(j, v_j)_{j \in J}\}$ to the cloud server. Finally, the blockchain performs pre-auditing calculations based on the integrity validation auxiliary information $IVA$: $\lambda = H_3(PID)^{\sum_{j=j_1}^{j_m} v_j} \cdot \sum_{j=j_1}^{j_m} (\varphi_j)^{v_j}$ .
- $ProofGen \rightarrow (\mu, Sig)$ : Following the reception of the challenge message $chal = \{(j, v_j)_{j \in J}\}$, the algorithm entails the Cloud Service Provider (CSP) crafting an appropriate response message that substantiates the audit of integrity. First, it aggregates data blocks: $\eta = \alpha^{-1} \sum_{j=j_1}^{j_m} v_j F[j] + \beta$ . Then, it calculates the aggregated signature $Sig$ and finally sends the integrity audit proof response message $\{\eta, Sig\}$ to the CSP via the blockchain. $Sig = \prod_{j=j_1}^{j_m} Sig_j^{v_j}$ .
- $Verify(\eta, \beta, Sig, \lambda) \rightarrow \{0,1\}$ : The initiation of the algorithm's execution is conducted by the blockchain, which employs preliminary auditing computations $\lambda$ to ascertain the validity of the integrity verification equation upon receiving the $Sig$ integrity audit verification response from the Cloud Service Provider (CSP). The equation should be satisfied, and the Decision Variable (DV) is set to 1; conversely, if not, DV is set to 0.

### 4.2. *Correctness*

The verification of the accuracy of the aforementioned equation can be conducted in the following manner.

$$e(g, Sig) = e(g, \prod_{j=j_1}^{j_m} (R_j \cdot T_j)^{v_j})$$

$$= \left\{ \begin{array}{l} e(g, \prod_{j=j_1}^{j_m} \varphi_i \cdot H_3(PID))^{t \cdot v_j} \\ \cdot \prod_{j=j_1}^{j_m} (\omega^{F[i]} \cdot H_4(Fid \| PID \| i)^{SK_{PID}})^{v_j} \end{array} \right\}$$

$$= \left\{ \begin{array}{l} e(Y, H_3(PID)^{\sum_{j=j_1}^{j_m} v_j} \cdot \prod_{j=j_1}^{j_m} (\sigma^j \cdot \kappa^j \cdot \gamma)^{v_j}) \\ \cdot e(g^{s \cdot H_2(PID)} \cdot g^{x \cdot H_2(PID)}, \\ \omega^{\sum_{j=j_1}^{j_m} v_j \cdot F[i]} \cdot \prod_{j=j_1}^{j_m} (H_4(Fid \| PID \| i))^{v_j}) \end{array} \right\}$$

$$= \left\{ \begin{array}{l} e(Y, \lambda) \cdot e({PID_a}^{H_2(PID)} \cdot mpk^{H_2(PID)}, \\ \omega^{\eta} \cdot \prod_{j=j_1}^{j_m} (H_4(Fid \| PID \| i))^{v_j}) \end{array} \right\}$$

## 5. Performance analysis

### 5.1. *Security proof*

This manuscript proceeds to conduct a security evaluation of a blockchain-based anonymous data integrity auditing scheme for the Internet of Things (IoT), denoted as BADIA. The evaluation encompasses an examination of data storage accuracy and the safeguarding of users' conditional identity privacy.

**Theorem.** Supporting conditional identity anonymization for cloud storage data integrity validation schemes ensures conditional identity privacy for users.

Proof. First, we prove the privacy of the user's outside identity. In our scheme, $RID \in \{0,1\}^{\rho}$ is the unique real identity of each user and $PID = (PID_a, PID_b)$ is the pseudo-identity generated by PKG, where $PID_a = g^s$, and PIDa and g are elements in the multiplicative cyclic group. In polynomial time, due to the discrete logarithmic difficulty problem, the adversary can calculate s negligibly after obtaining the $PID_a$ and $g$.

Furthermore, the Private Key Generator (PKG) leverages the master private key, denoted as x, to calculate the Partially Decrypted Identity ( $PID_b = ID \oplus H_1({PID_a}^x \| mpk \| Time)$ ). Within the cryptosystem, PKG is regarded as a reliable entity, safeguarding the master private key x from unauthorized access, particularly by adversaries. Owing to the complexity associated with the Computational Diffie-Hellman (CDH) problem, the adversary's acquisition of components $PID_a = g^s$ and $mpk = g^x$ does not enable the computation of $PID_a^x = g^{sx}$, and consequently, the actual recovery of the user's Real Identity ( $RID \in \{0,1\}^{\rho}$ ) remains unattainable.

### 5.2. Experiment

We analyze and compare the performance of this paper's scheme with related schemes in terms of integrity verification. For ease of description, the operational time frame for specific operations is shown below.

In this paper, four schemes are analyzed and compared in a test simulation. The test simulation operation is performed under Windows 10, using an Intel(R)i5-4210 2.40GHZ processor and 4GB of RAM, and the jpbc-2.0.0 cryptographic library as the test environment. The computation overhead time of all algorithms was calculated using C language and MIRACL, a cryptographic algorithm base library with version number 5.6.2.

The comparison of experimental simulation results shows that the proposed scheme in this paper is more lightweight compared to the existing schemes in data integrity verification.

The specific comparison results are shown in Figure 1.

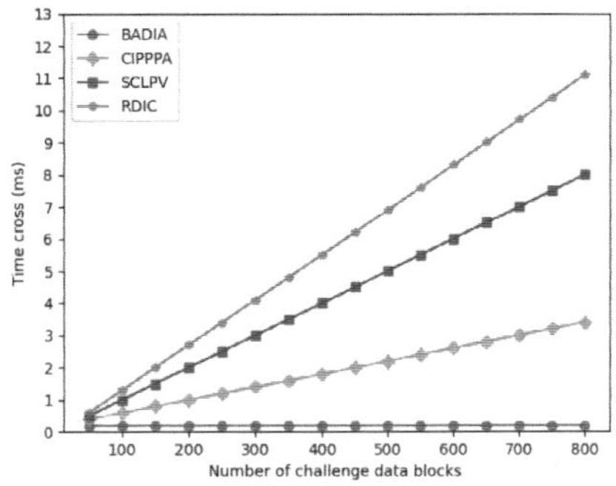


**Figure 1.** Comparison of communication overhead.

In the integrity verification phase of the methodology presented herein, the computational burden associated with blockchain (BC) verification exhibits consistent behavior, showing little to no variation with the expansion in the volume of challenge data blocks. Conversely, the alternative three frameworks display a linear relationship between the BC verification duration and the quantity of challenge data blocks, leading to increased computational demands on the BC, particularly when processing extensive block quantities. Consequently, our proposed method showcases enhanced computational efficiency, a characteristic substantiated by the data depicted in Figure 2.

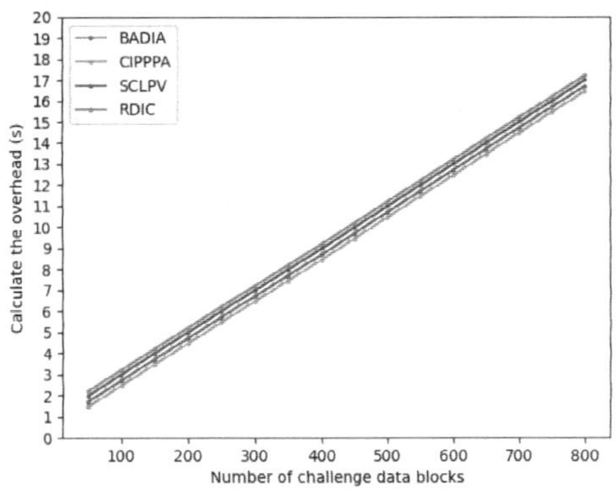


**Figure 2.** Experimental results of calculating expenses.

### 5.3. Conclusion

In response to the dual challenges of preserving user identity confidentiality and maintaining the integrity of data outsourced from IoT devices, this study introduces a blockchain-empowered scheme for anonymous data integrity auditing in cloud settings. Employing blockchain technology, the scheme conducts audits to ascertain data integrity, thereby reducing the potential threat posed by malevolent third-party auditors to the data's sanctity. An innovative feature of this system is its use of an identity-based aggregate signature algorithm, which facilitates streamlined and secure integrity checks. Crafted on the foundation of identity-based cryptography, this scheme eliminates the complexities inherent in the management of public key infrastructures, safeguarding user identity privacy and integrating mechanisms for the identification, exposure, and accountability of nefarious entities within the network. Through rigorous security assessments and performance benchmarks, it is validated that this scheme offers a reliable and effective solution for IoT device networks.

## References

[1] H. Qi, H. Li, W. Yu, and C. Wang, "IoV Edge Data Integrity Audit Method Based on Blockchain," 2023 International Conference on Blockchain Technology and Information Security(ICBCTIS), Xi'an, China, 2023, pp. 180-186, doi: 10.1109/ICBCTIS59921.2023. 00035.

[2] B. Zou, F. Nie, L. Xiao, T. Zhang, and C. Zhu, "MeDA: Using Blockchain for Patient-Controlled Medical Data Auditing in Institutions," 2022 22nd IEEE International Symposium on Cluster, Cloud and Internet Computing (CCGrid), Taormina, Italy, 2022, pp. 297-306, doi: 10.1109/CCGrid54584.2022.00039.

[3] J. Liu, H. Huang, Q. Chen, and Z. Huang, "A decentralized data integrity auditing scheme for cloud storage," 2023 International Conference on Data Security and Privacy Protection (DSPP), Xi'an, China, 2023, pp. 190-198, doi: 10.1109/DSPP58763.2023.10404960.

[4] Jianhong Zhang, Jie Wei, Jianan Zhang, Cao Bo, CKAA: Certificateless key-agreement authentication scheme in digital twin telemedicine environment, https://doi.org/10.1002/ett.4922.2023.10.27.

[5] J. Zhang and Q. Dong, Efficient ID-based public auditing for the outsourced data in cloud storage, Inf. Sci, doi. 343-344, no. 20, pp.1C14, 2016. 2000, pp. 2245–2249.

[6] Dean S, Klein J A. PSN Audits Uncover Mechanical Integrity Issues Part 3 of our series on audit findings focuses on common failings in mechanical integrity programs[J].Chemical Processing, 2022(4):84.

[7] R. Behl, A. Singh and A. Sinha, "A Study on Cloud-based data integrity of outsourced data using Blockchain," 2022 5th International Conference on Contemporary Computing and Informatics (IC3I), Uttar Pradesh, India, 2022, pp. 1193-1198, doi: 10.1109/IC3I56241.2022.10073235.

[8] ZHANG Xiaojun, ZHAO Jie, XU Chunxiang, et al. Conditional identity privacy-preserving public auditing for cloud-based WBANs against malicious auditors[J]. IEEE Transactions on Cloud Computing, 2021,[16]9(4): 1362C1375. doi: 10.1109/TCC.2019.2927219.

[9] YU Yong, AU M H, ATENIESE G, et al. Identity-based remote data integrity checking with perfect data privacy-preserving for cloud storage[J]. IEEE Transactions on Information Forensics and Security,2017, 12(4): 767C778. doi: 10.1109/TIFS.2016.2615853.

[10] J. Zhang, J. Zhang, Z. Lv, and R. Zhang, "On the Security of Privacy-preserving Data Aggregation Scheme Based on CRT in Mobile Crowdsensing System," 2023 International Conference on Cyber-Enabled Distributed Computing and Knowledge Discovery (CyberC), Jiangsu, China, 2023, pp. 331-334, doi: 10.1109/CyberC58899.2023.00058.

Journal of Physics: Conference Series **2813** (2024) 012014 doi:10.1088/1742-6596/2813/1/012014

# Improving chinese cross-modal retrieval with multi-modal transportation data

**Yanzhong He[1], Yanjiao Zhang[2] and Lin Zhu[2*]**

[1] School of Computer Science and Artificial Intelligence, Wuhan University of Technology, Wuhan, China
[2] School of Computer, Wuhan Qingchuan University, Wuhan, China

[*] Corresponding author's e-mail: linzhu_cs@163.com

**Abstract.** As societal development progresses and individual travel needs evolve, the demand for acquiring multi-modal transportation status information has steadily increased. The present domain of transportation status encompasses a wealth of multi-modal information, including vehicle trajectory data, traffic condition visual imagery, and textual information. Acquiring multi-modal transportation status information facilitates a rapid understanding of the prevailing traffic conditions at a given location. In this study, we investigate multi-modal transportation status data encompassing trajectories, textual descriptions, and visual imagery. We establish a multi-modal dataset comprising circuit, text, and image data and enhance the state-of-the-art cross-modal retrieval model within the Chinese domain. We utilize hash codes to compute the similarity between images and textual descriptions. During model evaluation using our collected dataset, the Chinese-CLIP-Hash model (our model) demonstrates performance improvements over the Chinese-CLIP model. Specifically, it exhibits a 1.0% improvement in R@10 for the Text-to-Image task and a 0.4% improvement in R@5 for the Image-to-Text task, along with a 0.2% enhancement in R@10. Furthermore, during the cross-modal retrieval process, it achieves a notable speed increase, exhibiting an 11% improvement in speed compared to the Chinese-CLIP model when handling queries from a set of 400 textual descriptions.

## 1. Introduction

With the advancement of society and the enhancement of individuals' quality of life, the issue of personal transportation has become a growing concern for many. There is an increasing demand among people for information on multi-modal traffic conditions. Currently, the transportation domain encompasses various data modalities such as trajectories, textual data, and images, each of which reflects real-time traffic conditions. To acquire information about traffic conditions, it is necessary to integrate data from multiple modalities to query the traffic status of a specific area. Despite the widespread use of cross-modal datasets in research, these datasets primarily focus on image and text annotations without catering to the domain of traffic conditions.

With the rise of large-scale language models in natural language processing (NLP), various large pre-trained models have emerged in the computer vision (CV) domain, consequently driving the development of vision-text cross-modal pretraining models. Current large models [1] have demonstrated remarkable capabilities, achieving state-of-the-art performance in various visual-language downstream tasks such as visual question answering [11], image captioning [12], and cross-modal retrieval[10]. This underscores the potential of models trained on extensive data for cross-modal learning [6]. However,

Journal of Physics: Conference Series **2813** (2024) 012014 doi:10.1088/1742-6596/2813/1/012014

due to the absence of cross-modal datasets designed explicitly for traffic conditions, current large models need to be equipped to address issues related to traffic conditions.

This paper makes three primary contributions: 1. The establishment of the world's first cross-modal dataset encompassing trajectories, text, and images collected from Didi ride-sharing courses in Chengdu, China, Weibo traffic police textual data, and static panoramic road images spanning from November 1st, 2016, to November 30th, 2016. 2. The improvement of the CLIP model for cross-modal hash retrieval. On our custom dataset, the Chinese-CLIP-Hash model outperforms the Chinese-CLIP model, achieving a 1.0% improvement in R@10 for the Text-to-Image task, a 0.4% improvement in R@5 for the Image-to-Text task, and a 0.2% improvement in R@10. Moreover, the cross-modal retrieval process exhibits increased speed, with an 11% enhancement compared to the Chinese-CLIP model when handling queries from 400 textual descriptions.

In the upcoming sections, we will introduce the cross-modal hash retrieval method, followed by an explanation of the process involved in establishing the trajectory-text-image multi-modal dataset and a presentation of the experimental results of the enhanced cross-modal hash retrieval. Finally, we will review related work and summarize the paper.

## 2. Methodology

CLIP (Contrastive Language-Image Pretraining) [3] is a multi-modal deep learning model developed by OpenAI, equipped with language and image comprehension capabilities. Its fundamental concept involves embedding text and images into the same space for contrastive learning, enabling the model to comprehend the relationship between the two. CLIP comprises a bimodal architecture, employing Convolutional Neural Networks (CNN) as the image encoder and Transformers as the text encoder, with contrastive learning used for model training.

Chinese-CLIP (Chinese Contrastive Language-Image Pretraining) [4] is the Chinese version of CLIP developed by Alibaba, explicitly optimized for Chinese language and image processing. Chinese-CLIP possesses multi-modal capabilities and is proficient in understanding Chinese text and processing images. It undergoes training through contrastive learning to facilitate interactions between text and pictures in various tasks. The model inherits CLIP's process of extracting features from images and text and uses cosine similarity to calculate similarity, achieving state-of-the-art performance in cross-modal retrieval within the Chinese domain.

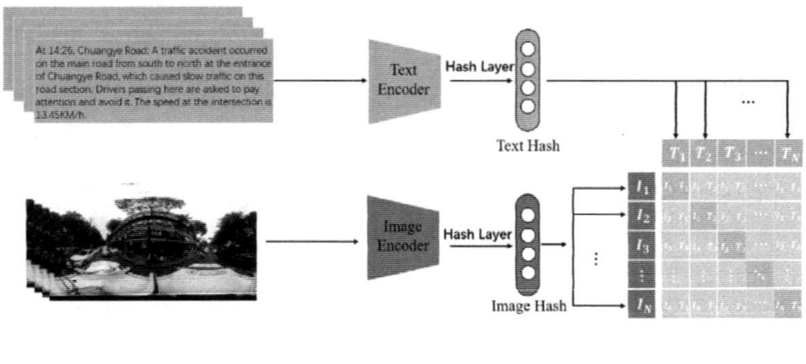


**Figure 1.** The framework of the Chinese-CLIP-Hash model.

In this paper, we extend the Chinese-CLIP model by combining it with autoencoders [8] and hashing [9] to generate fixed-length binary hash codes, enabling efficient similarity search and retrieval. This process involves feature encoding, where image or text features are encoded into low-dimensional continuous vectors using autoencoders, capturing the essential characteristics of the data. Subsequently, these constant vectors are transformed into fixed-length binary hash codes using hashing techniques, typically generated by applying hash functions to make similar data more likely to be equal in the hash space. Training can be carried out using supervised methods, utilizing label information to guide autoencoder learning for generating meaningful hash codes. Finally, the generated hash codes can be employed for efficient similarity search by calculating the Hamming distance of hash codes to identify similar data points rapidly. This model, through the learning of compact data representations and binary hash codes, reduces storage requirements and enhances retrieval efficiency. The Chinese-CLIP-Hash model architecture diagram is shown in Figure 1.

## 3. Method for establishing a multi-modal dataset

In this section, we will introduce the process of establishing multi-modal datasets.

### *3.1 Retrieval of traffic text events from Weibo*

Traffic events, constituting real-time traffic information, are predominantly disseminated through online media and broadcasts. Web scraping technology [2] conveniently extracts knowledge of information from online media sources. Web scrapers are proficient in automatically fetching data from the internet, which can then be saved in files or databases. In this study, we utilized the Selenium tool to automate web data collection and extraction. This tool simulates browser operations, enabling the retrieval of textual content from Weibo.

### *3.2 Named entity recognition for location and latitude-longitude information*

Numerous named entity recognition tools are widely available, with commonly used options such as HanLP [5] and NER [7]. HanLP, in particular, stands out as it facilitates text segmentation and exhibits excellent performance in identifying Chinese personal names, locations, and organizations, among other entities. In this paper, HanLP is employed to identify and extract location names within Weibo events.

Initially, the HanLP corpus was used for segmentation, but it yielded suboptimal results for toponyms due to their absence in the existing dictionary. To enhance accuracy, custom toponym mentions were included in a user-defined dictionary, eliminating the need to label noun parts of speech in documents. The custom dictionary was then integrated into the segmentation system, prioritizing its terms and reducing errors. Next, preprocessed microblog event files were read, and custom toponym recognition was improved to provide better segmentation results than the HanLP corpus. An 'ns' part-of-speech tag was assigned to toponyms, and these tagged words were merged into complete toponyms and saved to a file. For obtaining latitude and longitude coordinates of road segments, toponyms were used as starting and ending points. By assuming a straight line between these points and employing linear interpolation, coordinates for the entire congested road segment were calculated. Both Amap and Baidu Maps APIs were considered for address resolution. The Amap API, using the GCJ-02 coordinate system, was chosen for its consistency with the "Didi" dataset, which also employs GCJ-02. The geocoding API, accessible through an Amap account registration and API Key, transformed structured addresses into latitude and longitude coordinates and batch-saved them to files after extracting the data.

### *3.3 Retrieval of panoramic static images using Baidu Map api*

This paper presents an efficient method for retrieving panoramic static images through the Baidu Map API. The technique involves location recognition, coordinate conversion, and automated processing, enabling the acquisition of panoramic static images at specific locations. This capability supports the research on cross-modal hashing retrieval of trajectories, text, and pictures.

### *3.4 Preprocessing of trajectory data*

Trajectory data is gathered from GPS devices, offering high data update rates and accuracy. Data is collected every 3 seconds, capturing location coordinates. However, occasional interruptions or location inaccuracies may occur in areas with weak GPS signals. Consequently, preprocessing is required to address duplicates, errors, and data not meeting the specified requirements. The data cleaning phase involves the removal of trajectory data with time intervals of less than 6 seconds, latitude and longitude coordinates of 0 or outside specified ranges, and trajectory data with less than three minutes of order time. Subsequently, data processing includes converting timestamps to Beijing time format and filtering order data within the 6:00 AM to 6:00 PM time range to match the timing of Weibo events. Given the rich volume of order data, a compromise between accuracy and computational efficiency is necessary. This paper adopts a precision of four decimal places for latitude and longitude data. Utilizing the formula for calculating the distance between two points on Earth with latitude and longitude coordinates:

$$D_{AB} = R \cdot arcos(cos(Y_A) \cdot cos(Y_B) \cdot cos(X_A - X_B) + sin(Y_A) \cdot sin(Y_B)) \quad (1)$$

where A represents the coordinates $(X_A, Y_A)$, and B represents the coordinates$(X_B, Y_B)$, with R as the Earth's radius ($R = 6371.0km$). The precision adopted in this paper, accurate to four decimal places for latitude and longitude data, corresponds to a spatial variation of approximately 10 meters when latitude and longitude change by 0.0001. This level of precision is deemed suitable for everyday needs.

### *3.5 Connection of multi-modal data sets*

Trajectory speed information, Weibo textual information, and panoramic static images are connected based on timestamp and latitude-longitude matching, forming a multi-modal traffic condition information dataset encompassing 74 sets of trajectory-image-text multi-modal data, as well as 74×74 image-text matching pairs. The dataset includes three fields: image information, textual information fused from Weibo text and trajectory speed information, and matching labels (0 indicates no match, 1 indicates a match). The data is illustrated in Figure 2. below.

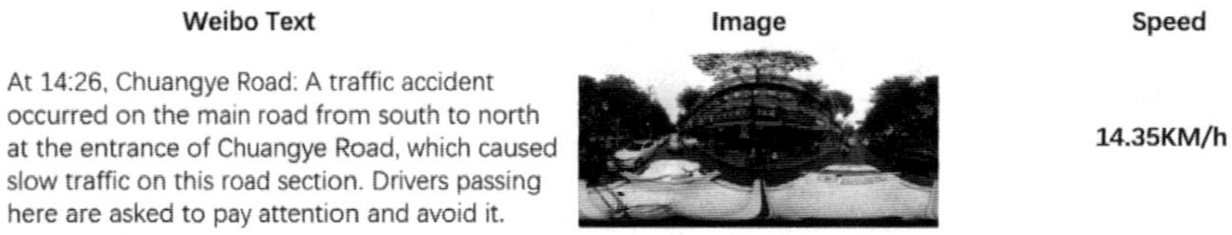


**Figure 2.** The details of the experimental data.

## 4. Experiment

In this section, we present a multifaceted examination of our study, encompassing data, experimental procedures, results, evaluation metrics, and the practical application of cross-modal retrieval.

### *4.1 Multi-modal traffic condition information dataset*

**Weibo Text Data:** It is statistically observed that 216 million people are active on Weibo daily, engaging in activities such as posting comments, sharing information, and staying updated with social dynamics. Therefore, acquiring traffic-related information from Weibo is an effective data source. In this study, we crawled data from the web version of Weibo, specifically from the official Weibo account of "Chengdu Traffic Police." This account predominantly publishes real-time traffic information for Chengdu. We collected Weibo content from November 1st to November 30th, 2016, which includes timestamps and the text of each Weibo post. **Trajectory Data:** The primary source of trajectory data in this paper is from Didi Chuxing, which offers comprehensive trajectory data for the Chengdu area to the academic community. The dataset covers Didi ride-hail order trajectories within Chengdu's Second Ring Road from November 1st to November 30th, 2016, spanning geographical coordinates [30.727818, 104.043333], [30.726490, 104.129076], [30.655191, 104.129591], and [30.652828, 104.042102]. Each trajectory data point includes driver IDs, order IDs, timestamps, longitude, and latitude. Didi Chuxing drivers often choose the fastest route based on long-term experience, considering traffic congestion conditions and the number of traffic lights at different times. Furthermore, the trajectory data reflects the vehicle's speed and the distance covered within fixed 3-second intervals, enabling the analysis of traffic conditions. Therefore, using Didi Chuxing's trajectory data provides a more accurate representation of the city's traffic conditions and offers valuable guidance to users for their journeys. **Image Data:** Baidu Maps panoramic static image data is sourced from Baidu Maps' Street View service, which provides real-time street view images for various locations. Baidu Maps static image data is derived from Baidu's street view collection vehicles, user-uploaded images, and data partnerships with collaborators. This data is processed and integrated, ultimately presented on the Baidu Maps application and website, providing users with an on-site street view experience. In this paper, Baidu Maps panoramic static images are obtained based on the latitude and longitude coordinates parsed from Weibo text, offering users an intuitive panoramic view of traffic conditions.

### *4.2 Experimental procedure*

We divided our collected dataset into training, testing, and validation sets with an 8:1:1 ratio. The enhanced Chinese-CLIP-Hash model was trained on the training set and tested on the testing set. Subsequently, the model was evaluated on the validation set. Both the CLIP model and Chinese-CLIP model were trained on their pre-trained models and assessed on the validation set.

### *4.3 Experimental results and evaluation metrics*

The evaluation metric used is Recall@K, where K = {1, 5, 10}. For all three models, we input the same testing set to compute similarity scores between images and text and return the top K most similar candidates. Recall@K measures the proportion of top K candidates that have matching data. Compared to the Chinese-CLIP model, Chinese-CLIP-Hash exhibited a 1.0% improvement in R@10 for the Text-to-Image task, a 0.4% improvement in R@5, and a 0.2% improvement in R@10 for the Image-to-Text task. Results are shown in Table 1. Furthermore, experimental results indicated that, when performing cross-modal retrieval on 400 text data points, CLIP took 1.0501 seconds, Chinese-CLIP took 1.0315 seconds, and Chinese-CLIP-Hash took 0.9166 seconds, resulting in an approximately 11% speed improvement.

**Table 1.** The details of the experimental result.

| Tasks | Text-to-Image | | | Image-to-Text | | |
|---|---|---|---|---|---|---|
| Metrics | R@1 | R@5 | R@10 | R@1 | R@5 | R@10 |
| CLIP | 48.3 | 65.1 | 70.2 | 50.2 | 67.8 | 71.2 |
| Chinese-CLIP | 50.6 | 68.6 | 78.3 | 50.5 | 68.7 | 79.2 |
| Chinese-CLIP-Hash | **50.5** | **68.5** | **79.3** | **50.3** | **69.1** | **79.4** |

### *4.4 Application of cross-modal retrieval*

The well-trained models were employed for cross-modal retrieval of text data (image data) to search for the top K most similar candidate image data (text data) within the database. Simultaneously, speed trajectory information was provided to facilitate cross-modal retrieval of traffic condition information.

## 5. Related work

**CLIP**, introduced by OpenAI in 2021, represents a robust multi-modal pre-trained model for image-text contrastive learning. Its prowess in cross-modal retrieval lies in its treatment of image-text pairs as cohesive entities. During training, the model's primary objective is to maximize the similarity between images and their corresponding text while minimizing the similarity between images and non-matching text. This process entails N image-text pairs, yielding N*N similarity scores for comparisons. In the inference phase, CLIP assesses multiple text prompts and an image, leveraging features from the Text Encoder and Image Encoder to compute similarity scores. These scores empower it to ascertain the most likely image category or to establish a correspondence between a single text prompt and a specific image within a set.

**Chinese-CLIP**, unveiled by Alibaba DAMO Academy in 2022, is an open-source multi-modal pre-trained model tailored to the Chinese language. Trained on a dataset comprising 200 million native Chinese image-text pairs, Chinese-CLIP enhances training efficiency and model performance through a two-stage training process. In the initial stage, the model initializes its dual-tower architecture by drawing upon pre-trained image and text models. During this phase, the image-side parameters remain frozen, establishing a crucial linkage between the language model and the existing image-pre-trained representation space. This step proves indispensable due to CLIP's text encoder initially being trained in English, necessitating retraining of the text representation model for the Chinese language. In the subsequent stage, the image-side parameters are unfrozen, enabling Chinese-CLIP to fine-tune native Chinese image and text data using contrastive learning. Within cross-modal retrieval, Chinese-CLIP calculates cosine similarity between image feature matrices and text feature matrices, thereby facilitating seamless cross-modal retrieval.

## 6. Conclusion

This paper investigates a cross-modal retrieval approach for traffic condition analysis, integrating trajectory data, Weibo text information, and panoramic static images into a multi-modal dataset. We enhance the state-of-the-art Chinese-CLIP model for cross-modal retrieval in the Chinese language domain by modifying the similarity computation method and utilizing autoencoders and hash codes to measure image-text similarity, resulting in an overall improvement. We compare the CLIP model trained with English text to Chinese-CLIP, finding that the Chinese-CLIP-Hash model outperforms Chinese-CLIP in the Text-to-Image task with a 1.0% increase in R@10 and the Image-to-Text task with a 0.4% increase in R@5 and a 0.2% increase in R@10. In the future, we aim to explore more efficient cross-modal retrieval methods to enhance the accuracy of traffic-related cross-modal retrieval.

**Acknowledgments**

The authors thank the National Innovation and Entrepreneurship Training Program for College Students (Grant No. 202310497058) for supporting this research.

## References

[1] Brown, Tom, et al. "Language models are few-shot learners." Advances in neural information processing systems 33 2020: 1877-1901.
[2] Heydon, Allan, and Marc Najork. "Mercator: A scalable, extensible web crawler." World Wide Web 2.4 1999: 219–229.
[3] Radford, Alec, et al. "Learning transferable visual models from natural language supervision." International conference on machine learning. PMLR, 2021.
[4] Yang, An, et al. "Chinese clip: Contrastive vision-language pretraining in chinese." arXiv preprint arXiv:2211.01335 2022.
[5] He, Han, and Jinho D. Choi. "The stem cell hypothesis: Dilemma behind multi-task learning with transformer encoders." arXiv preprint arXiv:2109.06939 2021.
[6] Li, Junnan, et al. "Blip-2: Bootstrapping language-image pre-training with frozen image encoders and large language models." arXiv preprint arXiv:2301.12597 2023.
[7] Liu, Wei, et al. "Lexicon enhanced Chinese sequence labeling using BERT adapter." arXiv preprint arXiv:2105.07148 2021.
[8] Kingma, Diederik P., and Max Welling. "An introduction to variational autoencoders." Foundations and Trends® in Machine Learning 12.4 (2019): 307-392.
[9] Liu, Song, et al. "Joint-modal distribution-based similarity hashing for large-scale unsupervised deep cross-modal retrieval." Proceedings of the 43rd international ACM SIGIR conference on research and development in information retrieval. 2020.
[10] Liu, Hao, et al. "Joint representation learning for multi-modal transportation recommendation." Proceedings of the AAAI Conference on Artificial Intelligence. Vol. 33. No. 01. 2019.
[11] Dua, Radhika, Sai Srinivas Kancheti, and Vineeth N. Balasubramanian. "Beyond vqa: Generating multi-word answers and rationales to visual questions." Proceedings of the IEEE/CVF Conference on Computer Vision and Pattern Recognition. 2021.
[12] Li, Hao, et al. "Uni-perceiver v2: A generalist model for large-scale vision and vision-language tasks." Proceedings of the IEEE/CVF Conference on Computer Vision and Pattern Recognition. 2023.

# Fractional order Lipschitz recurrent neural network with attention for long time series prediction

**Zelong Zhu, Chunna Zhao* and Yaqun Huang**

Faculty of Information Science and Engineering, Yunnan University, Kunming 650500, China

* Corresponding author's e-mail: zhaochunna@ynu.edu.cn

**Abstract**: Time series data prediction holds a significant importance in various applications. In this study, we specifically concentrate on long-time series data prediction. Recurrent Neural Networks are widely recognized as a fundamental neural network architecture for processing effectively time-series data. Recurrent Neural Network models encounter the gradient disappearance or gradient explosion challenge in long series data. To resolve the gradient problem and improve accuracy, the Fractional Order Lipschitz Recurrent Neural Network (FOLRNN) model is proposed to predict long time series in this paper. The proposed method uses the Lipschitz continuity to alleviate the gradient problem. The fractional order integration is applied to compute the hidden states of the Recurrent Neural Network in the proposed method. The intricate dynamics of long-time series data can be captured by fractional order calculus. It has more accurate predictions compared with Lipschitz Recurrent Neural Networks models. Then self-attention is used to improve feature representation. It can describe the correlation of features and improve predict performance. Some experiments show that the FOLRNN model achieves better results than other methods.

## 1. Introduction

Long-time series forecasting is very important in contemporary society. It is extensively applied in various domains including energy, economy, transportation, and climate. The accurate forecast results enable organizations and decision-makers to plan resources effectively.

In the past, mathematical and statistical methods served as the predominant approaches for time series forecasting. Those methods provide a solid theoretical foundation. However, those methods are often based on linearity and stationarity assumptions. Now, machine learning and deep learning models have gained prominence in time series forecasting.

Recurrent Neural Network (RNN) models have been widely acknowledged as suitable for processing sequential data [1]. However, the gradient problem is a significant challenge during the use of recurrent neural network processes and lengthy sequence data [2]. For the gradient problem, many approaches have been proposed to enhance the training stability and performance of recurrent neural networks [3].

RNNs system can be effectively described by using differential equations that are influenced by the input data [4]. Rubanova et al. introduced circular model equations and discrete integrals based on differential theory [5-6]. Erichson proposed the Lipschitz Recurrent Neural Network (LRNN), which reconstructs the recurrent matrix to mitigate the gradient problem [7]. LRNN effectively addresses gradient disappearance and explosion. It can get better results than previous recurrent neural networks. In a recurrent neural network dynamic system, differential equations are used to describe the dynamic influenced by the input data. The final hidden state of the dynamic system is obtained through one-order integration. Fractional order calculus is the extension of integer order calculus. Fractional order calculus models can describe the complex system more correctly. In this paper, fractional order integration is applied in the integration process to obtain an improved hidden state. Fractional Order Lipschitz Recurrent Neural Network (FOLRNN) is proposed based on fractional order integration. Our method can capture more complicated temporal dynamics and enhance the modeling capability of the hidden state in the system.

Self-attention mechanisms, originally popularized in NLP tasks, have been adapted and extended for time series analysis [8]. Wu et al. proposed Autoformer to improve the effect of the model by using inter-sequence attention and sequence decomposition [9]. Zhou et al. proposed FEDformer, which uses a Fourier transform to get the frequency domain information and uses a low-rank approximation to reduce the complexity of attention [10]. In this paper, self-attention is employed to capture feature correlations within the input data. Applying self-attention to capture feature interactions, the underlying relationships among the features can be described more correctly. It also may improve prediction accuracy.

In this paper, fractional order Lipschitz Recurrent Neural Network with Attention (FOLRNN-A) is proposed for long-time series prediction. The novelty of our work is summarized as follows:

1. We propose FOLRNN-A to handle long-time series prediction tasks. This model combines Recurrent Neural Network and self-attention to achieve accurate forecasts.
2. Benefiting from fractional order calculus in complex systems, fractional order integral is used to obtain better-hidden states in an RNN dynamic system.
3. Our model demonstrates competitive predictive results on five real-world datasets. The effectiveness of the model is illustrated through ablation experiments.

## 2. Model architecture

This section illustrates the framework of the model. Firstly, the fractional order Lipschitz Recurrent Neural Network module is described in detail. Then the self-attention module is introduced for feature fusion and enhancement. Lastly, the model's overall structure has been given.

### *2.1. FOLRNN module*

The training of recurrent neural networks has gradient problems. Erichson et al. use Lipschitz Recurrent Neural Networks (LRNN) to mitigate the gradient problems [7]. The dynamics of the system can be described by the differential equation. LRNN dynamics could be explained by Equation (1).

$$\begin{cases} \dot{h} = A_{\beta_A,\gamma_A} h + \tanh(W_{\beta_W,\gamma_W} h + Ux + b) \\ y = Dh \end{cases} \tag{1}$$

where A and W are hidden-to-hidden matrices calculated in the same way; h is the hidden state containing the past information; $\dot{h}$ is the time derivative; U is the input to the hidden encoder matrix; b is the offset.

In the LRNN model, the hidden state is obtained through the integration process. This integral is integer order then applied to the input data.

$$h = \Delta t \sum \dot{h}(t - r\Delta t) \tag{2}$$

The approach employed in this paper utilizes the Greenwall-Leitnikov (G-L) fractional order calculus for calculating the hidden states. Given the function f with respect to time $t$, the Greenwall-Leitnikov fractional calculus is defined as follows:

$$ {}_{t_0}^{GL}\mathscr{D}_t^{\alpha} f(t) = \lim_{h \to 0} \frac{1}{h^{\alpha}} \sum_{j=0}^{[(t-t_0)/h]} (-1)^j \binom{\alpha}{j} f(t - jh) \tag{3}$$

where $f$ is the integrated or differentiated function, $h$ is step length, and $\alpha$ denotes the calculus order, which is differential when $\alpha > 0$ and integral when $\alpha < 0$. It is important to note that in this equation, the order $\alpha$ is not necessarily an integer, which differs from integer calculus. G-L fractional calculus is derived from integer order differentiation, and the dynamics of recurrent neural networks are described by differential equations, so it is natural to describe the system dynamics through G-L fractional calculus. Equation (2) describes how to obtain the hidden state of the system by numerical integration, and the $\dot{h}$ in the equation is automatically calculated by the network. In this paper, the fractional integral is used to calculate the hidden state. Fractional order calculus serves as a natural extension of integer order calculus, offering increased flexibility in mathematical operations.

In our work, the hidden state h is calculated by using fractional integral as Equation (4):

$$D_t^{-p}(h(t)) = \Delta t^p \sum -1^r \binom{p}{r} h(t - r\Delta t) \tag{4}$$

where $D_t^{-p}$ represents the integral operator of the order p, in which p can be a non-integer value. The required coefficients $(-1)^r \binom{p}{r}$ can be computed before training.

*2.2 Attention module*

The self-attention can effectively capture correlations between inputs. It is based on computing the similarity and weighted scores. Some approaches involve predicting outcomes by computing correlations among time series data over time. However, when dealing with long time series tasks, this computation becomes more complex. The computational complexity of self-attention is quadratic with input length, which complicates the processing of long-series inputs. In this work, self-attention is used to compute correlations among different features, thereby fusing the features and representing them to a better one.

In the process of dealing with time series data, people find that interesting information resides in the patterns of change. This paper applies a differencing operation to data to enable the attention module to focus on the change trends within the series. The formula for the attention computation is presented in Equation (5).

$$\begin{aligned} Q,K,V &= W_{Q,K,V}(diff(X)) + b. \\ Attention(Q,K,V) &= Soft\max\left(\frac{QK^T}{\sqrt{d_Q}}\right)V. \end{aligned} \tag{5}$$

*2.3. Model architecture*

The model architecture proposed in this paper is shown in Figure 1. Input has been put into both FOLRNN and Attention modules. An attention module is followed by a FOLRNN module. The output of FOLRNN after the attention module is further processed by two linear layers, among which one contributes to the final prediction, while another output and input are added as new input. This updated input is then used for the subsequent attention module in the model.

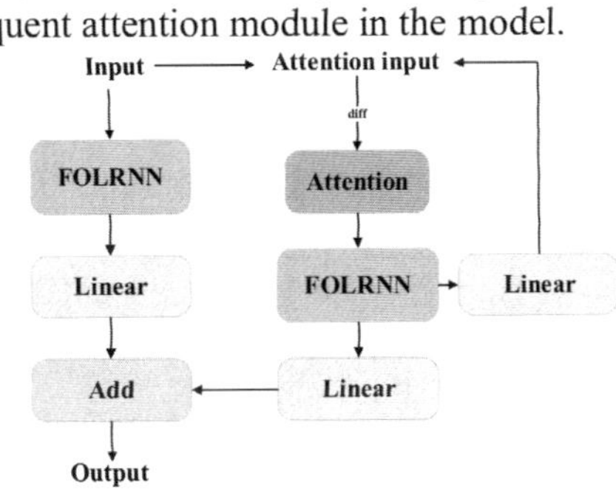


**Figure 1**. The structure of the FOLRNN-A model.

To improve both the generalization and accuracy of the model, the final prediction result is obtained by summing the outputs of all FOLRNN modules. Instead of employing iterative single-step prediction, this approach of multi-step prediction is utilized to alleviate the accumulation of errors over time. Through the adoption of multi-step prediction, the model seeks to enhance its overall performance and uphold robustness in prediction accuracy.

## 3. Experiment

This study evaluates the proposed model by assessing its performance using five real-world datasets that cover diverse domains, including energy, economy, transportation, and weather. Representative of time series data, these datasets form the foundation for assessing the performance and effectiveness of the model across various application areas.

**Table 1**. The predicted result on five datasets.

| Models | | **FOLRNN-A** | | FEDformer | | Autoformer | | Informer | | LSTNet | | LSTM | | TCN | |
|---|---|---|---|---|---|---|---|---|---|---|---|---|---|---|---|
| Metric | | MSE | MAE | MSE | MAE | MSE | MAE | MSE | MAE | MSE | MAE | MSE | MAE | MSE | MAE |
| Electricity | 96 | **0.158** | **0.253** | 0.183 | 0.297 | 0.201 | 0.317 | 0.274 | 0.368 | 0.680 | 0.645 | 0.375 | 0.437 | 0.985 | 0.813 |
| | 192 | **0.169** | **0.264** | 0.195 | 0.308 | 0.222 | 0.334 | 0.296 | 0.386 | 0.725 | 0.676 | 0.442 | 0.473 | 0.996 | 0.821 |
| | 336 | **0.186** | **0.282** | 0.212 | 0.313 | 0.231 | 0.338 | 0.300 | 0.394 | 0.828 | 0.727 | 0.439 | 0.473 | 1.000 | 0.824 |
| | 720 | **0.223** | **0.316** | 0.231 | 0.343 | 0.254 | 0.361 | 0.373 | 0.439 | 0.957 | 0.811 | 0.980 | 0.814 | 1.438 | 0.784 |
| Ettm2 | 96 | **0.195** | **0.282** | 0.203 | 0.287 | 0.255 | 0.339 | 0.365 | 0.453 | 3.142 | 1.365 | 2.041 | 1.073 | 3.041 | 1.330 |
| | 192 | 0.281 | 0.345 | **0.269** | **0.328** | 0.281 | 0.340 | 0.533 | 0.563 | 3.154 | 1.369 | 2.249 | 1.112 | 3.072 | 1.339 |
| | 336 | 0.351 | 0.391 | **0.325** | **0.366** | 0.339 | 0.372 | 1.363 | 0.887 | 3.160 | 1.369 | 2.568 | 1.238 | 3.105 | 1.348 |
| | 720 | 0.536 | 0.491 | **0.421** | **0.415** | 0.422 | 0.419 | 3.379 | 1.338 | 3.171 | 1.368 | 2.720 | 1.287 | 3.135 | 1.354 |
| Exchange | 96 | **0.120** | **0.258** | 0.139 | 0.276 | 0.197 | 0.323 | 0.847 | 0.752 | 1.551 | 1.058 | 1.453 | 1.049 | 3.004 | 1.432 |
| | 192 | 0.305 | 0.409 | **0.256** | **0.369** | 0.300 | 0.369 | 1.204 | 0.895 | 1.477 | 1.028 | 1.846 | 1.179 | 3.048 | 1.444 |
| | 336 | **0.408** | 0.489 | 0.426 | **0.464** | 0.509 | 0.524 | 1.672 | 1.036 | 1.507 | 1.031 | 2.136 | 1.231 | 3.113 | 1.459 |
| | 720 | 1.121 | **0.789** | **1.090** | 0.800 | 1.447 | 0.941 | 2.478 | 1.310 | 2.285 | 1.243 | 2.984 | 1.427 | 3.150 | 1.458 |
| Traffic | 96 | **0.529** | **0.305** | 0.562 | 0.349 | 0.613 | 0.388 | 0.719 | 0.391 | 1.107 | 0.685 | 0.843 | 0.453 | 1.438 | 0.784 |
| | 192 | **0.542** | **0.315** | 0.562 | 0.346 | 0.616 | 0.382 | 0.696 | 0.379 | 1.157 | 0.706 | 0.847 | 0.453 | 1.463 | 0.794 |
| | 336 | **0.556** | **0.323** | 0.570 | **0.323** | 0.622 | 0.337 | 0.777 | 0.420 | 1.216 | 0.730 | 0.853 | 0.455 | 1.476 | 0.799 |
| | 720 | 0.606 | **0.336** | **0.596** | 0.368 | 0.660 | 0.408 | 0.864 | 0.472 | 1.481 | 0.805 | 1.500 | 0.805 | 1.499 | 0.804 |
| Weather | 96 | **0.169** | **0.228** | 0.217 | 0.286 | 0.266 | 0.336 | 0.300 | 0.384 | 0.594 | 0.587 | 0.369 | 0.406 | 0.615 | 0.589 |
| | 192 | **0.213** | **0.274** | 0.276 | 0.336 | 0.307 | 0.367 | 0.598 | 0.544 | 0.560 | 0.565 | 0.416 | 0.435 | 0.629 | 0.600 |
| | 336 | **0.260** | **0.307** | 0.339 | 0.380 | 0.359 | 0.395 | 0.578 | 0.523 | 0.597 | 0.587 | 0.455 | 0.454 | 0.639 | 0.608 |
| | 720 | **0.334** | **0.363** | 0.403 | 0.428 | 0.419 | 0.428 | 1.059 | 0.741 | 0.618 | 0.599 | 0.535 | 0.520 | 0.639 | 0.610 |
| Count | | **29** | | 12 | | 0 | | 0 | | 0 | | 0 | | 0 | |

### 3.1 *Brief description of the datasets*

The "Electricity" dataset comprises hourly electricity consumption data for 321 customers, including the period from 2012 to 2014. The "ETT" dataset comprises voltmeter data recorded from July 2016 to July 2018. The "Weather" dataset consists of 10-minute records of 21 weather indicators for the entirety of 2020. The "Exchange" dataset contains data from eight countries, including the years from 1990 to 2016. Lastly, the "Traffic" dataset includes hourly road occupancy rates measured by various sensors situated on highways in the San Francisco Bay Area. For the ETT dataset, the data division into training, test, and validation sets follow a ratio of 6:2:2. For the remaining datasets, data is partitioned into training, test, and validation sets by using a ratio of 7:1:2.

### 3.2 *Implementation details*

In the conducted experiments, the L2 loss function is employed during training, and the batch size is set as 32. The optimization algorithm used is ADAM, with an initial learning rate of $10^{\wedge-2}$. After two epochs of training, the learning rate decays at a rate of 0.9. To prevent overfitting, an early stopping mechanism is implemented, and training is concluded within 20 epochs. All experiments are repeated five times and implemented with PyTorch. Training is performed on a single NVIDIA TITAN V 12 GB GPU. The LRNN hyperparameters $\beta$ and $\gamma$ are set to 0.7 and 0.01, respectively. The fractional order $\alpha$ is chosen to be 1.8 based on experimental comparisons. The number of attention modules in the model is either 1 (for datasets with fewer variables) or 2 (for datasets with more variables). The reasoning behind this decision is that, for datasets with a greater number of variables, the attention module can aggregate more pertinent information, leading to more accurate predictions.

### *3.3 Main results*

To evaluate the model's performance in multivariate long-time series prediction, a fixed input length of 96 has been used. The evaluation was carried out on output windows of varying lengths, specifically 96, 192, 336, and 720. The results of these experiments are presented in Table 1. This analysis allows for a comprehensive comparison of the model's predictive capabilities across different output window sizes. Our method exhibits superior performance compared to most baseline models across the five datasets, except for FEDformer. Furthermore, in most experiments, our method attains higher accuracy compared to FEDformer. Our method achieves a reduction of approximately 10% in Mean Squared Error (MSE) and Mean Absolute Error (MAE) on the Electricity dataset, and around 20% improvement on each prediction length of the Weather dataset. These improvements highlight the effectiveness of our method in enhancing prediction accuracy on these particular datasets.

### *3.4 Order of fractional order integration*

The optimal fractional order has consistently posed a challenge in the fractional realm, and a better order is obtained through the use of grid search. The search space chosen is (0, 2), which is a commonly utilized interval for fractional order studies. A set of fractional orders 0.2, 0.5, 0.8, 1.0, 1.2, 1.5, and 1.8 was chosen for evaluation.

Table 2. The results in a different ordershows our model predicts results from different orders of the integration. The results indicate that an increase in the order of fractional integration enhances the model's predictive performance.

**Table 2.** The results in a different order.

| | Alpha | 0.2 | | 0.5 | | 0.8 | | 1 | | 1.2 | | 1.5 | | 1.8 | |
|---|---|---|---|---|---|---|---|---|---|---|---|---|---|---|---|
| | Metric | MSE | MAE | MSE | MAE | MSE | MAE | MSE | MAE | MSE | MAE | MSE | MAE | MSE | MAE |
| Electricity | 96 | 0.176 | 0.268 | 0.170 | 0.263 | 0.168 | 0.259 | 0.168 | 0.261 | 0.161 | 0.255 | 0.160 | 0.254 | **0.158** | **0.253** |
| | 192 | 0.183 | 0.276 | 0.176 | 0.270 | 0.174 | 0.267 | 0.175 | 0.268 | 0.174 | 0.266 | 0.169 | 0.264 | **0.169** | **0.264** |
| | 336 | 0.199 | 0.294 | 0.195 | 0.289 | 0.191 | 0.285 | 0.190 | 0.285 | 0.187 | 0.282 | 0.189 | 0.283 | **0.186** | **0.282** |
| | 720 | 0.239 | 0.327 | 0.224 | 0.314 | 0.226 | 0.318 | 0.226 | 0.316 | **0.220** | **0.314** | 0.225 | 0.318 | 0.223 | 0.316 |
| Weather | 96 | 0.185 | 0.245 | 0.184 | 0.245 | 0.170 | 0.231 | 0.170 | 0.228 | 0.172 | 0.233 | **0.167** | **0.224** | 0.169 | 0.228 |
| | 192 | 0.227 | 0.284 | 0.213 | 0.268 | 0.215 | 0.273 | **0.209** | **0.268** | 0.220 | 0.276 | 0.210 | 0.271 | 0.213 | 0.274 |
| | 336 | 0.268 | 0.315 | 0.264 | 0.308 | 0.263 | 0.310 | 0.266 | 0.315 | 0.267 | 0.314 | **0.260** | **0.306** | **0.260** | 0.307 |
| | 720 | 0.340 | 0.376 | 0.339 | 0.366 | 0.336 | 0.363 | 0.333 | 0.360 | 0.340 | 0.379 | 0.341 | 0.378 | **0.334** | **0.363** |

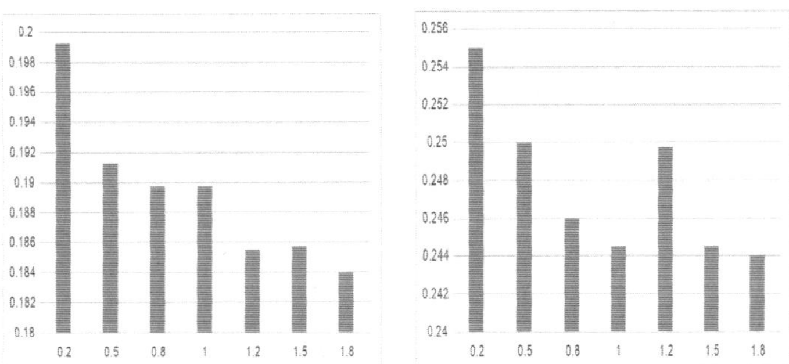


**(a)** Electricity **(b)** Weather

**Figure 2.** Different-order MSE result.

Journal of Physics: Conference Series **2813** (2024) 012015 doi:10.1088/1742-6596/2813/1/012015

Figure 2 shows the different order MSE results on two datasets. It is evident that in most cases, the MSE decreases as the fractional order increases. Higher fractional orders tend to yield more accurate results compared to lower fractional orders.

## 4. Summary

The FOLRNN model with attention is proposed for predicting long-time series data. It addresses the gradient problem typically encountered in RNNs by incorporating LRNN. This method can ensure more stable training. The prediction accuracy is enhanced by fractional order integration in LRNN for calculating the hidden states. To further enhance the performance of the model, a self-attention mechanism is used in the proposed method. It enables the model to focus on relevant features and capture correlations. This method can improve overall performance.

Our approach can capture complex dynamics and non-local dependencies and improve prediction accuracy compared with traditional integer-order integration. FOLRNN with Attention shows great performance in long-term series prediction. Its training and inference speed could be further improved.

## Acknowledgment

This research was funded by the National Natural Science Foundation of China (Grant numbers: 61862062 and 61104035).

## References

[1] Hinton G. E. Editor learning distributed representations of concepts. Proceedings of the eighth annual conference of the Cognitive Science Society, 1986, Amherst, MA.

[2] Bengio Y., Simard P., and Frasconi P. Learning of long-term dependencies with difficult gradient descent. IEEE transactions on neural networks, 1994, 5 (2): 157-66.

[3] Hochreiter S. and Schmidhuber J. Long short-term memory. Neural Computation, 1997, 9 (8): 1, 735-80.

[4] Funahashi K. I. and Nakamura Y. Approximation of dynamical systems by continuous time recurrent neural networks. Neural networks, 1993, 6 (6): 801-6.

[5] Rubanova Y., Chen R. T., and Duvenaud D. K. Latent ordinary differential equations for irregularly-sampled time series. Advances in neural information processing systems, 2019, 32.

[6] De Brouwer E., Simm J., Arany A., and Moreau Y. GRU-ODE-Bayes: Continuous modeling of sporadically-observed time series. Advances in neural information processing systems, 2019, 32.

[7] Erichson N. B., Azencot O., Queiruga A., Hodgkinson L., and Mahoney M. W. Lipschitz Recurrent Neural Networks. International Conference on Learning Representations, 2020.

[8] Vaswani A., Shazeer N., Parmar N., Uszkoreit J., Jones L., Gomez A. N., et al. Attention is all you need. Advances in neural information processing systems, 2017, 30.

[9] Wu H., Xu J., Wang J., and Long M. Autoformer: Decomposition transformers with auto-correlation for long-term series forecasting. Advances in Neural Information Processing Systems, 2021, 34: 22, 419-30.

[10] Zhou T., Ma Z., Wen Q., Wang X., Sun L., and Jin R. Fedformer: Frequency-enhanced decomposed transformer for long-term series forecasting. International Conference on Machine Learning, 2022: PMLR.

Journal of Physics: Conference Series **2813** (2024) 012016 doi:10.1088/1742-6596/2813/1/012016

# Research on the integrated evaluation model of database based on LDA and entropy weight method

**Junyun Mao, Hao Wen and Jing Huang***

Wuhan Railway Vocational College of Technology, Wuhan, Hubei, 430205, China

*Corresponding author's e-mail: hj5092000@126.com

**Abstract**: With more emphasis placed on the transferable acquisition and sustainable development of knowledge and data, the comprehensiveness, accuracy, and scientificity of databases are further called for. However, how to evaluate the effectiveness of the database needs further research. Thus, this paper builds an integrated evaluation model adopting the latent Dirichlet allocation model and entropy weight method. First, we calculate the contents of the database evaluation index system by the latent Dirichlet allocation model, which integrates the core orientation of evaluation with the characteristics of the target database. Second, the entropy weight method is used to ensure the weights of each index. Third, through empirical research, the reliability of the model is fully verified and suggestions to improve the database quality are specifically provided. The integrated evaluation model of a database is more comprehensive, accurate, and user-oriented, which provides effective reference for better evaluation of database quality.

## 1. Introduction

With the development of modernization, the comprehensive reform in the field of information technology has entered a critical stage, where more emphasis is placed on the transferable acquisition and sustainable development of knowledge and data, and a comprehensive model of database is further called for. In this background, the importance of a user-friendliness database that provides comprehensive, interdisciplinary, and application-oriented information resources has become a consensus. Meanwhile, to judge the effect of the various databases, the evaluation should be user-oriented and break existing discipline boundaries. However, the current research and practice on the evaluation mainly focuses on the cognition of relevant concepts, construction, and development status, and lacks practical research on data mining and text mining. For example, in [5], the present situation of resource database construction is analyzed, using the methods of concept definition, Likelihood Ratio test, and burst word analysis, but no corresponding evaluation method is proposed. In [6], the evaluation of the database lacks empirical research. In [7], the design of evaluation indexes needs more information analysis. How to carry out the database evaluation scientifically and reasonably needs further research.

Based on this, this paper calculates the contents of the database evaluation index system by the latent Dirichlet allocation model ("LDA"), which integrates the core orientation of evaluation with the characteristics of the target database. Based on this, the entropy weight method is used to ensure the weights of each index. Thus, through empirical research, this paper builds an integration evaluation model of a database that is more comprehensive, accurate, and user-oriented. It provides an effective reference for better evaluation of database quality.

## 2. Integrated evaluation model construction

### *2.1. Date mining and cleaning*

The content of database-evaluation indexes should include two parts: on the one hand, it should include the general elements and consensus on the evaluation of the database, which can be summarized and extracted by searching related academic articles; on the other hand, it should present the characteristic elements according to the different types and themes of the database, which can be sorted out based on the information contained in the database. The forms of data information can mainly be divided into text, image, audio, and video, of which text data is the most common form of presentation. Thus, the model proposed in this paper mainly collects text data. The information collected from the above two aspects constitutes the basis for determining the evaluation index system.

To reduce data noise, ensure integrity, and improve data quality, the data collected needs to be further cleaned. As the data sources are mainly academic papers and resources in the database, data cleaning is relatively simple and mainly includes two aspects: one is the cleaning of missing values and outliers, and the other is the cleaning of format and content errors. For missing values and outliers, specific judgments can be made according to their importance and missing rate, and fields can be completed, removed, or filled. For format and content errors, format inconsistency should be a concern. In this way, a solid data foundation for constructing the index system is laid.

### *2.2. Index system determination based on LDA*

Latent Dirichlet allocation, a kind of unsupervised learning technique, is a topic model that presents the topic of a document set as a probability distribution [2]. The main idea of LDA is that a document is characterized by serial topics, and a topic is composed of a set of certain words in the document, so the frequency of each word in a document can partially indicate the document's topics [8], thus enabling the reflection of the document's topics through modeling and calculating the most frequently occurring phrases.

We adopt the LDA topic model to analyze the data information obtained in 2.1., and calculate the key elements of the index system for the target database. The index system has two advantages: on the one hand, it contains the core orientation of the academic computer research field to the evaluation of the database, and avoids directional bias in evaluation; On the other hand, it takes the characteristics of the target database into full consideration and truly realizes the accuracy of the evaluation.

### *2.3. Integrated evaluation model of the database based on the entropy weight method*

The entropy weight method is an objective weighting method proposed by Shannon C. E [1]. According to the basic principles of information theory, entropy and information are a set of concepts with equal absolute values but opposite symbols [3]. When the entropy value is smaller, the data contains more useful information and the information value is bigger. When the entropy value is bigger, the data contains less useful information and the information value is smaller. In practice, the entropy weight of each index can be calculated respectively according to the information value of the indexes by the entropy weight method, and the index weights obtained are relatively objective.

The integration evaluation model of the database is built based on the entropy weight method. Firstly, we calculate the weight of each index obtained in 2.2 by the entropy weight method. Then, an evaluation model is formed, in which the weights of indexes are not affected by subjective factors. Thirdly, we calculate the comprehensive result which reflects the effect of the target database.

The calculation steps are as follows:

Firstly, it is assumed that the number of objects is $m$, the number of indexes is $n$, and an initial data matrix $\mathbf{X}$ is constructed, that is:

$$\mathbf{X} = \begin{bmatrix} x_{11} & \cdots\cdots & x_{1n} \\ x_{21} & \cdots\cdots & x_{2n} \\ x_{m1} & \cdots\cdots & x_{mn} \end{bmatrix} \tag{1}$$

Journal of Physics: Conference Series **2813** (2024) 012016 doi:10.1088/1742-6596/2813/1/012016

where $x_{ij}$ represents the value under the $j$-$th$ index of the $i$-$th$ object.

Then, the initial data matrix is standardized to eliminate the influence of dimension and value range differences between indexes, so that indexes with different value ranges can be compared and processed. After the isometric process, the matrix **R** is obtained.

$$\mathbf{R} = \begin{bmatrix} r_{11} & \ldots\ldots & r_{1n} \\ r_{21} & \ldots\ldots & r_{2n} \\ r_{m1} & \ldots\ldots & r_{mn} \end{bmatrix} \tag{2}$$

where $r_{ij} = x_{ij} / \sum_{i=1}^{m} x_{ij}, j=1,2,3,...,n$ .

The uncertainty of the $j$-$th$ index of the $i$-$th$ object is defined as $p_{ij}$, thus:

$$p_{ij} = r_{ij} / \sum_{i=1}^{m} r_{ij}, j = 1,2,...,n \tag{3}$$

Then, the information entropy of the evaluation is defined as $E$, thus:

$$\left.\begin{aligned} E_j &= -k \times \sum_{i=1}^{m} p_{ij} \ln p_{ij}, j = 1,2,..., n \\ k &= 1/\ln m, k>0 \end{aligned}\right\} \tag{4}$$

Finally, the entropy weight of each index is defined as $\omega$, and the weight of each index is calculated by information entropy, thus:

$$\omega_j = (1-E_j) / \sum_{j=1}^{n} (1-E_j), (\omega_j \in [0,1], \sum_{j=1}^{n} \omega_j = 1) \tag{5}$$

## 3. Example analysis and application

### *3.1. Index system determination of target research sample*

This paper takes Wuhan T College as the research sample to carry out case analysis, which is a full-time higher vocational institute specialized in railway programs. T College is a construction unit of high-level vocational colleges and programs in China, one of China's exemplary vocational colleges, and is a typical study sample.

According to the method mentioned in 2.2, the key elements of evaluation indexes of T College databases can be obtained, which is shown in Figure 1.

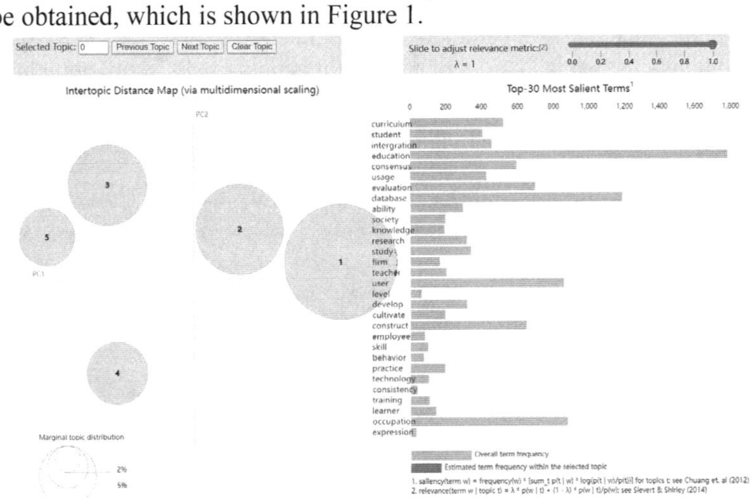


**Figure 1**. The key elements of evaluation indexes.

Journal of Physics: Conference Series **2813** (2024) 012016 doi:10.1088/1742-6596/2813/1/012016

Following the principles of systematization and hierarchy, the evaluation index system is constructed based on referring to the key topics, consulting education experts in related fields and soliciting the opinions of teachers and students [4][9]. The index system consists of a target layer (A), a primary indicator layer (B), and a secondary indicator layer (C). There are 5 indexes in the primary layer and 11 indexes in the secondary layer. The specific indicator system is shown in Table 1.

**Table 1.** The evaluation index system.

| Target layer (A) | Primary index layer (B) | Secondary index layer (C) |
|---|---|---|
| The effect of the target database A | Multiple and effective measures in obtaining Information B1 | The skills and accuracy of the data presentation C1 |
| | | The effects of theoretical information in general and professional contents C2 |
| | Scientific and reasonable management B2 | The effects of management when integrating the general and professional information C3 |
| | | The distribution of general and professional data information C4 |
| | Equal emphasis on theory and practice in training program data B3 | The guiding effects of practical activities in general and professional information C5 |
| | | The design of data distribution in different curricula C6 |
| | | Reasonable assessment of the user's progress C7 |
| | Complementary interdisciplinary information B4 | Abundant and natural presentation of general information elements in professional contents C8 |
| | | Abundant and natural presentation of professional information elements in general contents C9 |
| | Innovated conception and well-equipped facilities B5 | Correct and innovative understanding of the operation C10 |
| | | Rich allocation of relevant facilities C11 |

### *3.2. Questionnaire design and descriptive statistics*

Students are the inevitable participants, intuitive feelers, and ultimate beneficiaries of the database usage in this school, and the higher the students' satisfaction with the database is, the better the effect of integrating general education with professional education will be. So, students in this school are the objects of the questionnaire survey.

Since students in all grades have free access to the databases of this college, questionnaires are distributed to all students randomly. A total of 1, 900 questionnaires are distributed in the survey, 1, 705 questionnaires are recovered, and 1, 653 valid questionnaires are obtained, with an effective recovery rate of 87%. The first part of the questionnaire is to collect the basic information of the respondents. The second part is the respondents' evaluation of the school's database, including 11 indexes, which are shown in Table 1 and provide the main data source for the construction of the model mentioned in 2.3. The third part is a survey of the need for improving the integration of different kinds of databases.

The students' evaluations are divided into five levels, which are "very satisfied/ very good presentation", "relatively satisfied/ better presentation", "generally satisfied/ general presentation", "not very satisfied/ not very good presentation", and "very dissatisfied/ very bad presentation". Since the entropy weight method is mainly applied to evaluation systems with quantitative evaluation indexes, this paper semantically quantifies and calculates the evaluation grade assignment. The constructed comment set D={d1, d2, d3, d4, d5} corresponds to five evaluation levels, and the specific classification is shown in Table 2.

**Table 2.** The overall information of samples.

| Comment set D | d1 | d2 | d3 | d4 | d5 |
|---|---|---|---|---|---|
| Evaluation | Very satisfied/ Very good presentation | Relatively satisfied/ Better presentation | Generally satisfied/ General presentation | Not very satisfied/ Not a very good presentation | Very dissatisfied/ Very bad presentation |
| Value | 5 | 4 | 3 | 2 | 1 |

*3.3. Model calculation*

The quantitative evaluation grades are brought into the constructed evaluation model of the database (seen in 2.3). Since the sample size is relatively large, which is 1, 653, only 10 samples are randomly selected for display. The data results are shown in Table 3.

**Table 3.** Data table of evaluation grade assignment of the database in T school (part).

| Serial number | Secondary index layer | | | | | | | | | | |
|---|---|---|---|---|---|---|---|---|---|---|---|
| | C1 | C2 | C3 | C4 | C5 | C6 | C7 | C8 | C9 | C10 | C11 |
| 1 | 4 | 4 | 4 | 4 | 4 | 4 | 4 | 4 | 4 | 4 | 3 |
| 2 | 3 | 3 | 3 | 3 | 3 | 3 | 3 | 4 | 4 | 3 | 4 |
| 3 | 3 | 3 | 3 | 3 | 3 | 3 | 3 | 4 | 4 | 3 | 3 |
| 4 | 3 | 3 | 3 | 3 | 3 | 3 | 3 | 4 | 4 | 3 | 3 |
| 5 | 4 | 4 | 4 | 4 | 4 | 4 | 4 | 4 | 4 | 4 | 4 |
| 6 | 3 | 3 | 3 | 3 | 3 | 3 | 3 | 4 | 4 | 3 | 3 |
| 7 | 3 | 3 | 3 | 3 | 3 | 3 | 3 | 4 | 4 | 3 | 3 |
| 8 | 3 | 3 | 3 | 3 | 3 | 3 | 3 | 4 | 4 | 3 | 4 |
| 9 | 4 | 4 | 4 | 4 | 3 | 4 | 4 | 5 | 1 | 1 | 3 |
| 10 | 4 | 4 | 4 | 4 | 4 | 4 | 4 | 4 | 4 | 4 | 4 |

According to Formula (2), the table data are standardized, and the results are shown in Table 4.

**Table 4.** Standardized data (part).

| Serial number | Secondary index layer | | | | | | | | | | |
|---|---|---|---|---|---|---|---|---|---|---|---|
| | C1 | C2 | C3 | C4 | C5 | C6 | C7 | C8 | C9 | C10 | C11 |
| 1 | 1.1059 | 1.0987 | 1.0978 | 1.0989 | 1.1003 | 1.1016 | 1.0974 | 0.8921 | 0.8862 | 1.1011 | 0.8148 |
| 2 | 0.8294 | 0.8240 | 0.8233 | 0.8242 | 0.8253 | 0.8262 | 0.8231 | 0.8921 | 0.8862 | 0.8258 | 1.0864 |
| 3 | 0.8294 | 0.8240 | 0.8233 | 0.8242 | 0.8253 | 0.8262 | 0.8231 | 0.8921 | 0.8862 | 0.8258 | 0.8148 |
| 4 | 0.8294 | 0.8240 | 0.8233 | 0.8242 | 0.8253 | 0.8262 | 0.8231 | 0.8921 | 0.8862 | 0.8258 | 0.8148 |
| 5 | 1.1059 | 1.0987 | 1.0978 | 1.0989 | 1.1003 | 1.1016 | 1.0974 | 0.8921 | 0.8862 | 1.1011 | 1.0864 |
| 6 | 0.8294 | 0.8240 | 0.8233 | 0.8242 | 0.8253 | 0.8262 | 0.8231 | 0.8921 | 0.8862 | 0.8258 | 0.8148 |
| 7 | 0.8294 | 0.8240 | 0.8233 | 0.8242 | 0.8253 | 0.8262 | 0.8231 | 0.8921 | 0.8862 | 0.8258 | 0.8148 |
| 8 | 0.8294 | 0.8240 | 0.8233 | 0.8242 | 0.8253 | 0.8262 | 0.8231 | 0.8921 | 0.8862 | 0.8258 | 1.0864 |
| 9 | 1.1059 | 1.0987 | 1.0978 | 1.0989 | 0.8253 | 1.1016 | 1.0974 | 1.1151 | 0.2216 | 0.2753 | 0.8148 |
| 10 | 1.1059 | 1.0987 | 1.0978 | 1.0989 | 1.1003 | 1.1016 | 1.0974 | 0.8921 | 0.8862 | 1.1011 | 1.0864 |

The standardized data is substituted into Formula (3) and Formula (4) to obtain the information entropy of each index $E_j$. Then, the entropy value is substituted into Formula (5) to obtain the weight of each secondary index $\omega_j$, and the results are shown in Table 5.

**Table 5.** Information entropy and weight of secondary index layer.

| Secondary index layer | Information entropy $E$ | Weight $\omega$ |
|---|---|---|
| C1 | 9.96659E-01 | 0.09028 |
| C2 | 9.96653E-01 | 0.09042 |
| C3 | 9.96949E-01 | 0.08245 |
| C4 | 9.96983E-01 | 0.08153 |
| C5 | 9.96963E-01 | 0.08205 |
| C6 | 9.96988E-01 | 0.08138 |
| C7 | 9.96976E-01 | 0.08172 |

| Secondary index layer | Information entropy $E$ | Weight $\omega$ |
|---|---|---|
| C8 | 9.95355E-01 | 0.12552 |
| C9 | 9.95434E-01 | 0.12338 |
| C10 | 9.96962E-01 | 0.08209 |
| C11 | 9.97069E-01 | 0.07920 |

### *3.4. Consistency test*

To further verify the reliability of the integration evaluation model constructed in this paper, 10 experts from the fields of communications and signals, professional education, computer technology, and quality supervision were invited to anonymously evaluate the weights of 11 evaluation indexes. The results of the experts' evaluation are shown in Table 6.

**Table 6.** Experts' evaluation results.

| Serial | Weight of secondary index layer | | | | | | | | | | |
|---|---|---|---|---|---|---|---|---|---|---|---|
| number | C1 | C2 | C3 | C4 | C5 | C6 | C7 | C8 | C9 | C10 | C11 |
| 1 | 0.1000 | 0.1000 | 0.0700 | 0.0900 | 0.0800 | 0.0900 | 0.0800 | 0.1300 | 0.1100 | 0.0800 | 0.0700 |
| 2 | 0.1000 | 0.0950 | 0.0750 | 0.1000 | 0.0950 | 0.0800 | 0.0800 | 0.1200 | 0.1100 | 0.0700 | 0.0750 |
| 3 | 0.0900 | 0.0900 | 0.0800 | 0.0700 | 0.1000 | 0.0800 | 0.0800 | 0.1200 | 0.1400 | 0.0700 | 0.0800 |
| 4 | 0.0800 | 0.0800 | 0.0900 | 0.0700 | 0.0700 | 0.0700 | 0.0900 | 0.1300 | 0.1500 | 0.0800 | 0.0900 |
| 5 | 0.0900 | 0.0900 | 0.0800 | 0.0900 | 0.0900 | 0.0800 | 0.0800 | 0.1100 | 0.1200 | 0.0900 | 0.0800 |
| 6 | 0.0900 | 0.0900 | 0.0900 | 0.0800 | 0.0800 | 0.0900 | 0.0900 | 0.1100 | 0.1200 | 0.0800 | 0.0800 |
| 7 | 0.0900 | 0.0800 | 0.0800 | 0.0900 | 0.0900 | 0.0800 | 0.0800 | 0.1200 | 0.1300 | 0.0900 | 0.0700 |
| 8 | 0.0900 | 0.0900 | 0.0900 | 0.0800 | 0.0800 | 0.0900 | 0.0900 | 0.1300 | 0.1000 | 0.0800 | 0.0800 |
| 9 | 0.0900 | 0.0900 | 0.0900 | 0.0700 | 0.0700 | 0.0700 | 0.0700 | 0.1400 | 0.1300 | 0.0900 | 0.0900 |
| 10 | 0.0800 | 0.1000 | 0.0800 | 0.0700 | 0.0800 | 0.0800 | 0.0800 | 0.1400 | 0.1200 | 0.0900 | 0.0800 |

Comparing the weight of each index in the model with the 10 experts' evaluation results, we can obtain 11 groups of index weight data, and the Kendall coordination coefficient is used to test the internal consistency of the 11 groups of index weight data. SPSSAU 23.0 is used for statistical analysis [10], and the results are shown in Table 7.

**Table 7.** Kendall W coordination coefficient analysis.

| Number of index weight data groups | Number of indexes | Kendall $W$ coordination coefficient | $p$ |
|---|---|---|---|
| 11 | 11 | 0.616 | $<0.001$ |

It can be seen from the above table that the Kendall coordination coefficient of each index weight in the model and the weight of each index obtained by 10 experts is 0.616, which is between 0.6 and 0.8, showing significance$(p<0.001)$. The reliability of the model is fully verified.

### *3.5. Model application*

According to the integrated evaluation model of the database constructed in this paper and the data in Table 5, the scores of each index of T College are obtained. The weighted scores of each index of 1, 653 samples are calculated on average to obtain T school's final score of database evaluation, as shown in Table 8.

**Table 8.** Standardized data (part).

| Serial | Weighted value | | | | | | | | | | |
|---|---|---|---|---|---|---|---|---|---|---|---|
| number | C1 | C2 | C3 | C4 | C5 | C6 | C7 | C8 | C9 | C10 | C11 |
| 1 | 0.4514 | 0.4521 | 0.4122 | 0.4076 | 0.4102 | 0.4069 | 0.4086 | 0.5021 | 0.4935 | 0.4104 | 0.2970 |
| 2 | 0.3386 | 0.3391 | 0.3092 | 0.3057 | 0.3077 | 0.3052 | 0.3064 | 0.5021 | 0.4935 | 0.3078 | 0.3960 |
| 3 | 0.3386 | 0.3391 | 0.3092 | 0.3057 | 0.3077 | 0.3052 | 0.3064 | 0.5021 | 0.4935 | 0.3078 | 0.2970 |
| 4 | 0.3386 | 0.3391 | 0.3092 | 0.3057 | 0.3077 | 0.3052 | 0.3064 | 0.5021 | 0.4935 | 0.3078 | 0.2970 |
| 5 | 0.4514 | 0.4521 | 0.4122 | 0.4076 | 0.4102 | 0.4069 | 0.4086 | 0.5021 | 0.4935 | 0.4104 | 0.3960 |
| 6 | 0.3386 | 0.3391 | 0.3092 | 0.3057 | 0.3077 | 0.3052 | 0.3064 | 0.5021 | 0.4935 | 0.3078 | 0.2970 |
| 7 | 0.3386 | 0.3391 | 0.3092 | 0.3057 | 0.3077 | 0.3052 | 0.3064 | 0.5021 | 0.4935 | 0.3078 | 0.2970 |
| 8 | 0.3386 | 0.3391 | 0.3092 | 0.3057 | 0.3077 | 0.3052 | 0.3064 | 0.5021 | 0.4935 | 0.3078 | 0.3960 |

| Serial number | Weighted value | | | | | | | | | | |
|---|---|---|---|---|---|---|---|---|---|---|---|
| | C1 | C2 | C3 | C4 | C5 | C6 | C7 | C8 | C9 | C10 | C11 |
| 9 | 0.4514 | 0.4521 | 0.4122 | 0.4076 | 0.3077 | 0.4069 | 0.4086 | 0.6276 | 0.1234 | 0.1026 | 0.2970 |
| 10 | 0.4514 | 0.4521 | 0.4122 | 0.4076 | 0.4102 | 0.4069 | 0.4086 | 0.5021 | 0.4935 | 0.4104 | 0.3960 |
| Final score | 4.1731 | | | | | | | | | | |

It can be seen that T school's final score on database evaluation is 4.1731. According to Table 2, the students in this school are satisfied with the synthesis of information in the database, and the effect of the database is relatively good in this school. Compared with the weight ratio, the scores of indicators C11, C7, and C3 are relatively high. It means that the conception to continue to improve and innovate the construction of databases is deeply rooted in the faculties' and students' hearts, the assessments can effectively integrate general information with professional information, and the presentation of data information is also quite effective. On the contrary, the scores of indicators C8, C1, and C6 are relatively low. It means that the generalization of professional data information and the distribution of different types of information skills need to be further strengthened or improved in this school.

## 4. Conclusion

With the comprehensive information revolution entering the critical stage, integrated and multivariate databases have been widely explored and practiced, which take the innovative, composite, and applied usage of information technology as the core and break through the boundaries of existing knowledge disciplines. To evaluate the efficiency of the database more accurately, this paper builds an integrated evaluation model adopting the latent Dirichlet allocation model and entropy weight method which integrate the core orientation of evaluation with the characteristics of the target database. Through empirical research, this evaluation model proves to be objective, scientific, and effective and can provide a reference basis for further realizing the better evaluation of database quality.

## Acknowledgments

This paper is funded by the Hubei Provincial Education Science Planning Project "Research on the Optimization of the Spatial Layout of Higher Vocational Education in Hubei Province under the Regional Integration Development Strategy" (Project No. 2022GB195).

## References

[1] Shannon C. E. (1948). A Mathematical Theory of Communication. Bell System Technical Journal. 27, 379- 423. DOI: 10.1002/j.1538-7305.1948.tb01338.x.
[2] Blei D. M., Ng A. Y., and Jordan M. I. (2003). Latent Dirichlet Allocation. Journal of Machine Learning Research, 3, 993- 1022. DOI: 10.5555/944919.944937.
[3] Michael Mitzenmacher and Eli Upfal (2005). Probability and Computing. Cambridge University Press, Cambridge. DOI: http://dx.doi.org/10.1017/S0963548309990344.
[4] Bassiou N. K. and Kotropoulos C. L. (2014). Online PLSA: Batch updating techniques including out-of-vocabulary words. IEEE Transactions on Neural Networks and Learning Systems, 25, 1, 953-1, 966. DOI: 10.1109/TNNLS.2014.2299806.
[5] Ba S. and Zhang Q. (2023). Research Status and Development Strategies of Digital Education Resources in China. Software Guide, 22, 225-236. DOI: 10.11907/rjdk. 222007.
[6] Hua M. J., Hu Y., et al. (2023). The Study on the Construction of Digital Resources in Military Academies. Journal of Higher Education, 7, 60-63. DOI: 10.19980/j.cn23-1593/g4.2021.12.015.
[7] Lu T. and Le C. (2021). A subjective-and-objective comprehensive dynamic evaluation method of university educational cloud resources. Journal of Nanchang Institute of Technology, 40, 113-118. DOI: 10.3969/j.issn.1006-4869.2021.01.020.
[8] Yang Kai (2017). Exploring Entropy-based Term Weighting Schemes in Latent Dirichlet Allocation. DOI: CNKI: CDMD:2.1017.857211.

[9] Ji Zhang (2022). Research on the Construction of a Digital Resource Evaluation Index System in Higher Vocational Colleges. DOI: 10.27230/d.cnki.gnmsu.2022.001120.
[10] The SPSSAU Project (2023). SPSSAU (Version 23.0) [Online Application Software]. Retrieved from https://www.spssau.com.

# Improved image segmentation method based on Otsu thresholding and level set techniques

**Zhipeng Jing and Bo Tang***

School of Mathematics and Computing Sciences,Guangxi Colleges and Universities Key Laboratory of Data Analysis and Computation,Guilin University of Electronic Technology, Guilin, 541001, China

*Corresponding author's e-mail: apriltb@163.com

**Abstract:** Image segmentation is a complex task in the field of image processing and computer vision, as it faces challenges such as noise, low contrast, and intensity variations. Due to the simplicity, efficiency, and ease of implementation of the Otsu algorithm, as well as its ability to improve accuracy in complex scenarios and enhance robustness against noise and lighting variations, a new grayscale pixel probability selection method has been proposed based on the classic Otsu method and Chan-Vese level set. By applying the Bayesian probability formula, the probability of a grayscale pixel is determined by the global prior and non-local prior. This improved Otsu method is combined with the Chan-Vese level set curve evolution method, resulting in a better image segmentation approach. Comparative experiments have demonstrated the advantages of this method.

## 1. Introduction

In the era of 5G technology, images have become a crucial medium for humans to receive external information. As technology continues to advance rapidly, the significance of image information in our daily lives has grown tremendously. Starting from the 1960s, there has been a wave of research focused on image information processing technology, leading to remarkable advancements in the field. This technology is no longer limited to research but has found applications in various fields, including medical image diagnosis [1], autonomous driving [2], satellite image processing [3], environmental analysis [4], image search engines [5], and more.

The thresholding method is a popular technique in image segmentation due to its simplicity and efficiency. The classical global binarization algorithm known as the Otsu method was proposed by Otsu et al. [6] in 1979. However, this method has limitations in accurately segmenting images with low contrast between the target and background or non-uniform grayscale values of the target. These shortcomings make the method sensitive to noise, which can result in inaccurate segmentation. To address these limitations, researchers have made enhancements to the Otsu method. For example, Hou et al. [7] introduced the minimum variance thresholding method to identify and handle outliers in the data. Xu et al. [8] found that the Otsu method tends to select a threshold with higher variance when the intra-class variance of the two classes is unequal. To address this issue, they proposed an improved Otsu method that incorporates a constrained gray-level search range, aiming to enhance the accuracy and robustness of the Otsu method in image segmentation tasks. Additionally, Yang et al. [9] introduced an enhanced approach based on the median method, replacing the respective mean values in the image with the median values of the foreground and background. This method aims to determine the optimal threshold by maximizing the interclass variance.

Traditional image segmentation methods cannot meet the increasing demands for image segmentation. Deep neural networks have been widely studied as they can learn highly complex functions required for tasks. After 2000, with the improvement of hardware devices, deep learning methods have regained attention and demonstrated significant capabilities in various image processing tasks. Le et al. [10] proposed a horizontal level set model based on deep recurrent neural networks, which is more robust and efficient compared to traditional level sets. Hu et al. [11] utilized the level set function as a loss function in neural networks, enabling the network model to converge faster during training. Subsequently, Chen et al. [12] incorporated region and length constraints from the level set model into the loss function of neural networks, leading to improved segmentation performance. In recent years, the rapid development of deep learning methods and their inherent potential have made deep learning the primary choice for image segmentation tasks.

## 2. Improvement of models

In this section, we propose an improved image segmentation method that combines the Otsu method and the C-V method. This method utilizes Bayesian probability to select pixel values of the grayscale image and employs a genetic algorithm for pre-segmenting the target region. The maximum outer rectangle of the target region is used as the initial curve for the C-V method, which then segments the image with inhomogeneous intensity. By integrating these techniques, our approach aims to enhance segmentation accuracy, especially in cases of intensity inhomogeneity.

Assuming that the range of gray levels in a grayscale image is (0, 255), and the gray value of the pixel at each point in the image is f(x,y), then f(x,y)∈ {0,1,··· ,255}, and let the threshold value be T $0 \leq T \leq 255$, then

$$g(x,y)=\begin{cases}0 & f(x,y)\prec T\\ 1 & f(x,y)\geq T\end{cases} \tag{1}$$

In the given context,g(x,y) represents the value of each pixel point in the binarized image. If g(x,y) equals 1, it indicates that the point belongs to the foreground, while g(x,y)equaling 0 signifies that the point is part of the back ground. To ensure a consistent overall brightness across the image, the Retinex algorithm is initially employed. Following this, based on the previous Otsu theory, Bayesian probability is utilized to express the likelihood of an occurrence of a grayscale i. Initially, the proportion of pixels in the foreground region of the image compared to the entire image A is approximately calculated using a prior computation. Finally, the probability of the occurrence of grayscale i in the improved Otsu method can be determined.

$$\mathrm{p}_i=\frac{A\times B_i}{A\times B_i+(1-A)\times C_i} \tag{2}$$

$$\sum_{\mathrm{i}=0}^{L-1} p_i=1 \tag{3}$$

$$P_1(k)=\sum_{i=0}^{k} p_i \tag{4}$$

$$\mathrm{n}_k=\sum_{i=0}^{k} i p_i \tag{5}$$

In the given context, Bi represents the probability that a foreground pixel is identified as belonging to the foreground region. Similarly, Ci denotes the probability that a background pixel is incorrectly classified as part of the foreground region. The following expressions are obtained according to the Otsu method:

Journal of Physics: Conference Series **2813** (2024) 012017 doi:10.1088/1742-6596/2813/1/012017

$$\sigma_B^2(k) = P_1(k)[n_1(k) - n_C]^2 + P_2(k)[n_2(k) - n_C]^2 \tag{6}$$

$$\text{k}^* = \arg\max \sigma_B^2(k) \tag{7}$$

$$\text{n}_C = \sum_{i=0}^{L-1} ip_i = P_1(k) \cdot n_1(k) + P_2(k) \cdot n_2(k) \tag{8}$$

In the context provided, a genetic algorithm is employed to optimize the threshold value k* in the improved Otsu method, enhancing the accuracy of the segmentation process. This optimized threshold value is then utilized to obtain the foreground region Ω0. Subsequently, the maximum outer rectangle of the foreground region is selected as the closed curve C in the C-V method.

$$C(s,t) = (x(s,t), y(s,t)), t \geq 0 \tag{9}$$

$$\text{c}_1 = \frac{\int_\Omega I(x,y)H(\phi)dxdy}{\int_\Omega H(\phi)dxdy} \tag{10}$$

$$\text{c}_2 = \frac{\int_\Omega I(x,y)(1-H(\phi))dxdy}{\int_\Omega (1-H(\phi))dxdy} \tag{11}$$

$$E(c_1, c_2, C) = u\oint_C ds + \lambda_1 \iint_{\Omega_1} (I - c_1)^2 dxdy + \lambda_2 \iint_{\Omega_2} (I - c_2)^2 dxdy \tag{12}$$

$$\frac{\partial \phi}{\partial \text{t}} = \delta_\varepsilon [udiv(\frac{\nabla \phi}{|\nabla \phi|}) - \lambda_1 (I - c_1)^2 + \lambda_2 (I - c_2)^2] \tag{13}$$

The integration of the C-V method with the improved Otsu method enables continuous evolution of the segmentation process, leading to a more precise target region. This enhanced approach boosts the accuracy of gray pixel probability within the Otsu method and strengthens the resilience of the C-V method. Additionally, this combination accelerates the evolution process, yielding superior segmentation outcomes. Overall, this refined methodology is valuable for advanced image segmentation applications.

Algorithm The main algorithmic steps of the model in this paper are represented as follows:

Step 1: Even out the overall brightness of the input image using the Retinex algorithm;

Step 2: The Bayesian probability formula (2) was used to calculate the probability pi, i = 0,1,2,--- , L − 1 of gray scale pixel i;

Step 3: Using equation (4 ), for k = 0,1,2,---,L − 1, calculate P1(k);

Step 4: Using equation (5), fork = 0,1,2,--- ,L − 1, calculate nk;

Step 5: Using equation (8), calculate the global gray mean nC;

Step 6: Using equation (6), for k = 0,1,2,--- ,L − 1, calculate the interclass variance $\sigma_B^2(k)$

Step 7: Genetic algorithm was used to obtain the optimal Otsu threshold k*;

Step 8: The maximum outer rectangle of the pre-segmented target region is used as the initialization curve in the C-V method C;

Step 9: Using equations (10) and (11), calculate c1 and c2;

Step 10:Evolve the level set function according to the evolution equation (13) of the curve;

Step 11:Reinitialize the level set function so that it is once again a symbolic distance function;

Step 12:Determine whether the level set function reaches the convergence condition, if it reaches the convergence condition, then stop the evolution, otherwise return to step 9.

Journal of Physics: Conference Series **2813** (2024) 012017 doi:10.1088/1742-6596/2813/1/012017

## 3. Comparison and analysis of results

To ensure consistency in the experimental design, the parameters in the genetic algorithm for the threshold method are unified with a population size of 10, crossover probability set to 0.9, and mutation probability set to 0.2. The C-V method employs a time step of 0.1, and $\lambda 1$and$\lambda 2$values of 1, while the length regularization term parameter is set to 0.

In this section, Figure 1 shows the iterative process of gray uneven image segmentation. The level set curve is observed to evolve rapidly and achieve excellent segmentation results even when dealing with complex boundaries and uneven gray distribution.

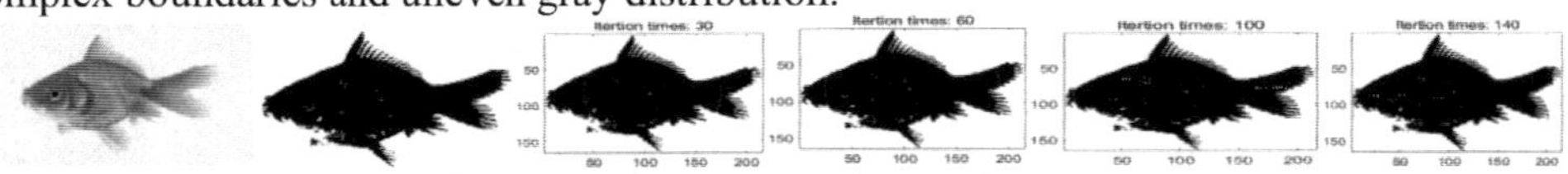

**Figure 1.** Segmentation of intensity inhomogeneous images.

To validate the proposed model's superiority over classical models, several image samples from the classical image segmentation library are selected. Segmentation comparisons were performed on these samples using GAC, CV (simplified M-S model), neural network-based U-shaped network (UNet), and our proposed model.

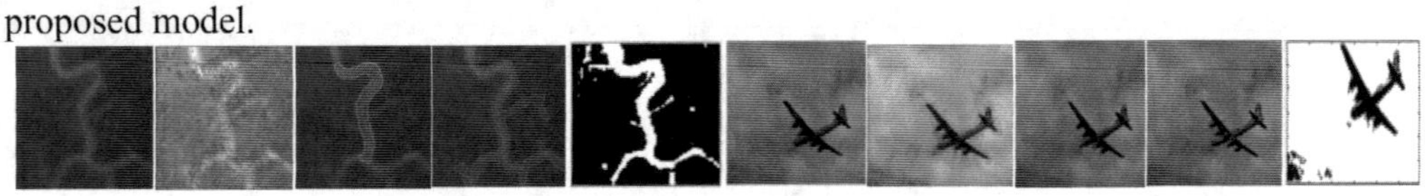

**Figure 2.** From left to right, the segmentation results of the initial outline, GAC, CV, UNet, and the model in this article are shown in sequence.

Based on the experimental results presented in Figure 2, our proposed model outperforms the UNet in terms of aircraft segmentation while also demonstrating superior performance in medical image segmentation. Furthermore, our model exhibits high accuracy and algorithmic efficiency.

To more directly demonstrate the segmentation accuracy between the improved model and the current mainstream U-shaped network (UNet), this article adopts the Dice Similarity Coefficient (DSC) to measure the accuracy of image segmentation.

(a) (b) (c) (d) (e)

**Figure 3.** Classic pictures split gallery original pictures.

**Table 1.** Similarity of dice for different model segmentations.

| Partition model | GAC | CV | UNET | Model of this paper |
|---|---|---|---|---|
| Figure 3.(a) | 0.31855 | 0.80134 | 0.85972 | 0.85016 |
| Figure 3.(b) | 0.23893 | 0.61688 | 0.81696 | 0.73327 |
| Figure 3.(c) | 0.27280 | 0.56443 | 0.73604 | 0.69197 |
| Figure 3.(d) | 0.34695 | 0.70802 | 0.82286 | 0.78991 |
| Figure 3.(e) | 0.07414 | 0.32566 | 0.56012 | 0.56713 |

From the similarity measures in Table 1, it can be observed that the model in this article outperforms GAC and CV in segmentation effectiveness. Although it does not perform as well as the currently popular neural network UNet in segmentation, the similarity is very close.

## 4. Conclusion

This paper introduces an innovative image segmentation model within the variational level set framework, merging an enhanced Otsu method with the C-V method. The model uses the Otsu method's optimal threshold from maximum inter-class variance and Bayesian probability for grayscale likelihood.

It also optimizes curve evolution by using the target region's maximum outer rectangle as the initial curve, overcoming the Otsu method's limitations in non-homogeneous grayscale. Comparatively, our method outperforms traditional models (GAC, M-S, CV) in managing intensity inhomogeneity, with segmentation results comparable to UNet.

**Acknowledgment**
The work was supported by the Guangxi Key Laboratory of Cryptography and Information Security (No.GCIS201928).

**References**

[1] Novikov A A, Lenis D, Major D, et al. Fully convolutional architectures for multi-class segmentation in chest radiographs[J]. IEEE transactions on medical imaging, 2018, 37(8): 1865-1876.
[2] Sagar A, Soundrapandiyan R K. Semantic segmentation with multi-scale spatial attention for self-driving cars[C]//Proceedings of the IEEE/CVF International Conference on Computer Vision. 2021: 2650-2656.
[3] Yoshihara A, Hascoet T, Takiguchi T, et al. Satellite image semantic segmentation using a fully convolutional network[J]. Red, 2018, 655: 690.
[4] King A, Bhandarkar S M, Hopkinson B M. A comparison of deep learning methods for semantic segmentation of coral reef survey images[C]//Proceedings of the I EEE conference on computer vision and pattern recognition workshops. 2018: 139 4-1402.
[5] Martinsson J, Mogren O. Semantic segmentation of fashion images using feature pyramid networks[C]//Proceedings of the IEEE/CVF International Conference on Computer Vision Workshops. 2019: 0-0.
[6] Otsu N. A threshold selection method from gray-level histograms[J]. IEEE transactions on systems, man, and cybernetics, 1979, 9(1): 62-66.
[7] Hou Z, Hu Q, Nowinski W L. On minimum variance thresholding[J]. Pattern Recognition Letters, 2006, 27(14): 1732-1743.
[8] Xu X, Xu S, Jin L, et al. Characteristic analysis of Otsu threshold and its applications[J]. Pattern recognition letters, 2011, 32(7): 956-961.
[9] Yang X, Shen X, Long J, et al. An improved median-based Otsu image thresholding algorithm[J]. AASRI procedia, 2012, 3: 468-473.
[10] Le T H N, Quach K G, Luu K, et al. Reformulating level sets as deep recurrent neural network approach to semantic segmentation[J]. IEEE Transactions on Image Processing, 2018, 27(5): 2393-2407.
[11] Chen X, Williams B M, Vallabhaneni S R, et al. Learning active contour models for medical image segmentation[C]//Proceedings of the IEEE/CVF conference on computer vision and pattern recognition. 2019: 11632-11640.
[12] Hu P, Shuai B, Liu J, et al. Deep level sets for salient object detection[C]//Proceedings of the IEEE conference on computer vision and pattern recognition. 2017: 2300-2309.

Journal of Physics: Conference Series **2813** (2024) 012018 doi:10.1088/1742-6596/2813/1/012018

# Research on regional difference mechanism of circulation innovation based on panel regression model

**Chongyan Li**[1,2] **and Xianke Li**[1,2*]

[1]Faculty of Humanities and Social Sciences, Macao Polytechnic University, Macao, 999078, China
[2]School of Economics and Management, Guangzhou College of Applied Science and Technology, Guangzhou, Guangdong, 511370, China

[*]Corresponding author's e-mail: lixianke678@163.com

**Abstract:** The panel regression model is remarkably persuasive in exploring the causal relationship. Therefore, combined with China's macroeconomic data from 2004 to 2021, this paper uses a panel regression model to reveal the influence mechanism of regional differences affecting innovation flow. Using the improved Theil index to construct panel data of regional differences, this paper explores the reasons for the unbalanced development of technological innovation circulation in China. The empirical results of the fixed effects model show that after controlling for the differences in economic opening environment and firm profit space, the differences in government support and circulation environment significantly affect the regional differences in technological innovation circulation level. This result helps to verify the existence of a "double wheel drive" mechanism in the construction of China's circulation innovation system, that is, the dual leading role of "effective government" and "efficient market" with Chinese characteristics.

## 1. Introduction

Regression models are generally used to explore causal relationships between variables. However, for the regression model of cross-section data or time series, the conclusion is not convincing enough because of the lack of data. Therefore, establishing a regression analysis of panel data can greatly improve its reliability. Using panel data to set up fixed effects or individual effects regression models, causal analysis of macroeconomic data has been widely used. This paper constructs panel data of regional differences by using the improved Theil index, makes an empirical analysis of the mechanism of differences in different regions, and verifies the "double wheel drive" mechanism in the construction of China's circulation innovation system.

It is noted that the existing research has carried out in-depth research on the path of enterprise technological innovation, but in-depth research on the macro-level technological innovation system is still relatively scarce. The outline of China's national 14th Five-Year Plan has defined the direction of development of China's circulation innovation system. However, the factors that drive the flow of technological innovation in the region should be noticed. Therefore, in Section 2, this paper introduces the main research methods and the structure of panel data. Section 3 discusses the factors that influence regional differences in the flow level of technological innovation and establishes a fixed effects model to explore the causal relationship by using panel data of regional differences after excluding the influence of regional economic development imbalance. In Section 4, the influential factors affecting the regional flow level of technological innovation are discussed, and then a robust analysis is carried out to further determine the influence mechanism. Section 5 summarizes.

## 2. Literature review

As for the research on circulation innovation systems, domestic scholars mostly focus on analyzing circulation innovation from the perspective of government guidance and enterprise motivation, while foreign scholars pay more attention to analyzing circulation innovation from the perspective of industrial chain and product coordination.

On the one hand, Liu (2022) and Chen et al. (2022) believed that effective government plays a significant role in the top-level design of circulation innovation [1-2]. Chen et al. (2023) believed that the central and local linkage to build a key core technology innovation ecosystem and long-term mechanism could effectively solve the “bottleneck” problem of China’s key core technologies [3]. Cui et al. (2021) held that the interaction between government and enterprise has a significant effect on the decision-making of circulation innovation of enterprises and has different effects on enterprises with different political backgrounds [4].

On the other hand, Chen & Yang (2021) believed that the framework of circulation innovation should be integrated from four-dimensional circulation processes, such as innovation system and policy, industrial chain, and innovation chain, the large, medium, and small enterprises and other innovation subjects, and innovation factors [5]. To strengthen the dominant position of enterprise innovation, the deep integration of IUR (Industry-University-Research) should be promoted. Attention should be given to the important role of entrepreneurs, and the circulation and creation of enterprises [6] should be promoted as the starting point. The upstream and downstream of the industrial chain can be integrated, product collaborative innovation [7-8], collaborative innovation [9], and the mechanism and principles of the innovation platform [10-11] can be promoted through the innovation chain, promoting the establishment of a circulation system for technological innovation. In this process, we should focus on the synergistic innovation effect of intellectual property protection on promoting circulation innovation [12], the link effect of knowledge sharing on circulation innovation [13], and the synergy effect of intellectual property protection on circulation innovation, the synergy effect of product demand and product supply on circulation innovation [7][14].

Therefore, there are two driving factors for the construction of a circulation innovation system: government-led and market-led. In addition, as the Chinese-style modernization enters a new stage of development, the dual role of “effective Government” and “efficient market” will play is the key to the success of the Chinese model.

## 3. Research design

### *3.1 Research area and data sources*

Based on the consistency of statistical caliber and the availability and completeness of data, this study selects 30 provinces and municipalities directly under the Central Government of China (excluding Tibet and Hong Kong, Macao, and Taiwan, due to the absence of some data) as research objects. The sample data come from 2005-2022 China Statistical Yearbook.

### *3.2 Improved theil index method*

As the traditional Theil index relies too much on the absolute value of a single variable, to eliminate the impact of unbalanced economic development in different regions as much as possible, the weight is reflected by the proportion of GDP of each region in the GDP of the whole country (or sub-region) in each observation year. The improved Theil index calculation formula is shown in Formulas (1-2):

$$TC_i = \frac{Y_i}{Y} \ln\left(\frac{Y_i/Y}{G_i/G}\right) \tag{1}$$

$$TC = \sum_{i=1}^{n}\left[\frac{Y_i}{Y} \ln\left(\frac{Y_i/Y}{G_i/G}\right)\right] \tag{2}$$

where, $TCI_i$ is the Theil index of co-innovation of the i province and city, and TCI is the Theil index of co-innovation of the whole country (sub-region), reflecting the overall imbalance of regional co-innovation level in China. $Y_i$ is the co-innovation level of the i province, and Y is the national co-innovation level. $G_i$ is the GDP of the i province and city, and G is the national GDP.

## 4. Regression analysis of regional technological innovation circulation differences

### *4.1 Variable analysis*

This paper adopts panel data from the Theil index of circulation innovation in various regions of China from 2004 to 2021 for regression analysis. Explained variables, explanatory variables, and control variables are described as follows:

The explained variable. Differences in circulation levels. This variable reflects the difference in the flow level of technological innovation in different regions. The Theil index of flow innovation is taken as the proxy variable and set as variable V11. The calculation steps are as follows: First, the entropy weight method is used to calculate the flow index of technological innovation. Hence, the variable V1 is set, and then the Theil index of flow innovation is calculated using Formula 2. The same is below.

The explanatory variables. It includes two indicators, namely, the difference in government support and the difference in circulation environment. First, the difference in government support reflects the difference in the intensity of the government's financial support. The government's financial support mainly involves providing direct or indirect financial support for technological innovation activities and the implementation of technological innovation, such as setting up various technological innovation awards, holding technical exchange conferences, and organizing technology patent exhibitions. This paper uses the total expenditure of local finance for science and technology to replace the government's support for the circulation of technological innovation. Therefore, the calculation procedure is to first calculate the government support level and set it as variable V2, and the government-supported Theil index is used as the proxy variable in this paper and is set as variable V12, by using Formula 2.

Second, the difference in circulation environment reflects the overall difference in the technology circulation environment in the market. The higher the technological innovation available for circulation is, the more enthusiastic the market response will be, and the more actively enterprises will apply for technological innovation patents. Since this paper studies the causal relationship, the success of technological innovation in the previous year is used to replace it. According to the data obtained, the calculation step is to first calculate the circulation environment and set it as variable V3, and the Theil index of circulation environment is used as the proxy variable in this paper and set as variable V13 by using Formula 2.

Control variables. There are many factors affecting the flow of technological innovation because technological innovation will be attached to services or products. This paper chooses the difference in economic opening environment and the difference in enterprise profit space as the control variables. The economic opening environment and set as variable V4. The economic opening environment difference was set as variable V14 with the economic opening environment Theil index as the proxy variable. The difference of enterprise profit space is set as variable V15 with the Theil index of enterprise profit space as the proxy variable V5.

### *4.2 Model establishment*

This paper uses the Hausman test to determine whether the regression model should be a fixed effect model or a random effect model. The results show that chi2=49.35 and the p-value is much less than 0.01, rejecting the null hypothesis that the individual effect of the omitted variables is not correlated with the explanatory variables. Therefore, a fixed effect model can be established to analyze the factors leading to regional differences, and the model of Formula 3 can be established.

$$V11_{it} = \alpha_t + \beta_1 V12_{it} + \beta_2 V13_{it} + \beta_3 V14_{it} + \beta_4 V15_{it} + \mu_{it} \quad (3)$$

where α is a constant term, β is the regression coefficient, μ is the random error term and follows the normal distribution, i is for region, and t stands for year.

*4.3 Regression results*

Using Stata16.0 software to execute the fixed effect regression command on panel data, the results are shown in Table 1. The regression results show that the overall goodness of fit of the model is 0.6441, and the explanatory power is good. The F-value is 10.69, and the overall model is significant at the 1% confidence level. Specifically, the explanatory variables V12, V13, V14, and V15 have passed the significance test at the confidence level of 1% and 5%, respectively, indicating good model results. The regression coefficients of the difference in government support and the difference in circulation environment are 0.1090 and 0.1942 respectively, which have a great impact on the regional differences in the circulation of technological innovation. The regression coefficients of economic opening environment difference and enterprise profit space difference are 0.0143 and 0.0309, respectively, which have little influence on the regional differences of technological innovation circulation. The constant term is 0.0056, which passes the significance test at the 1% confidence level. The gap between the regression coefficients of the explanatory and control variables tends to be reasonable. This shows that the regional circulation innovation system is greatly affected by government support. And it also proves that market utilization has a significant impact on the regional circulation innovation system.

**Table 1.** Regression results of Theil index of regional circulation innovation.

| | | | | | |
|---|---|---|---|---|---|
| | R-sq: | | Obs. per group: | | |
| | within = 0.0779 | | min = 18 | | |
| | between = 0.8835 | | avg = 8.0 | | |
| | overall = 0.6441 | | max = 18 | | |
| | | | F (4, 506) =10.69 | | |
| | Corr (u_i, Xb) = 0.7778 | | Prob > F = 0.0000 | | |
| | Variable | Coef. | Std. Err. | t | P>t |
| Explanatory variables | V12 | 0.1090 | 0.0274 | 3.98 | 0.000*** |
| | V13 | 0.1942 | 0.0663 | 2.93 | 0.004*** |
| Control variables | V14 | 0.0143 | 0.0042 | 3.39 | 0.001*** |
| | V15 | 0.0309 | 0.0123 | 2.51 | 0.012** |
| | _cons | 0.0056 | 0.0011 | 4.93 | 0.000*** |
| F test that all u_i=0: F (29, 506) = 6.05 | | | Prob > F = 0.0000 | | |

Note: *, **, and *** indicate significance at 10%, 5%, and 1% levels, respectively. The same is below.

## 5. Robust analysis

*5.1 Model establishment*

To scientifically and reasonably analyze the factors that contribute to the formation of regional differences in circulation innovation, a robust analysis is carried out by using the variable data before calculating the Theil index. Because the Theil index results in this paper have excluded the influence of regional economic development imbalance, a new explanatory variable, namely economic growth, is added based on the above analysis, which is set as variable V6.

This paper uses the Hausman test to determine whether the regression model for robust analysis should be a fixed effect model or a random effect model. The results show that chi2=312.76 and the p-value is much less than 0.01, rejecting the null hypothesis that the individual effect of the omitted variables is not correlated with the explanatory variables. Therefore, a fixed effect model can be established to conduct a robust analysis of the factors leading to regional differences, and the model of Formula 4 can be established.

$$V1_{it} = \alpha_t + \beta_1 V2_{it} + \beta_2 V3_{it} + \beta_3 V4_{it} + \beta_4 V5_{it} + \beta_5 V6_{it} + \mu_{it} \quad (4)$$

where α is a constant term, β is the regression coefficient, μ is the random error term and follows the normal distribution, i is for region, and t stands for year.

*5.2 Regression results*

Since explained variable V1, explanatory variables V2 and V3, and control variables V4 and V5 have all been set as ratio variables, regression analysis can be conducted. The newly added explanatory variable V6 also needs to be logarithms to make the data as stable as possible. Because the results of the Hausman test show that a fixed effect model should be established, Stata16.0 software is used to execute the fixed effect regression command with the region as a cluster, and the results are shown in Table 2.

The regression results show that the overall goodness of fit of the model is 0.4340, indicating that the explanatory power is acceptable. The F value is 181.16, and the model is significant overall at a 1% confidence level. This robust model has good persuasion. Specifically, the p-values of all variables pass the significance test at the 1% confidence level, and the model results are good. The regression coefficients of explanatory variables V2, V3, and V6 are 0.8324, 0.1037, and 0.059, respectively, which are all positive, indicating that government support, circulation environment, and economic growth have a positive impact on the circulation level of regional technological innovation, and the strength of the impact decreases in turn. The control variables V4 and V5 are 0.0006 and -0.0015, respectively, which have a small impact on regional differences in circulation innovation. The constant term is -0.5933, which passes the significance test at the confidence level of 1%. The results of robust regression once again show that the "double wheel drive" of the government and the market is the key to the regional circulation innovation system.

**Table 2.** Regression results of fixed effect regression with region clustering.

| | | | | | |
|---|---|---|---|---|---|
| | R-sq: | | Obs. per group: | | |
| | within = 0.8111 | | min = 18 | | |
| | between = 0.8111 | | avg = 18.0 | | |
| | overall = 0.4340 | | max = 18 | | |
| | | | F (5,29) = 181.16 | | |
| | Corr (u _ i, Xb) = -0.7336 | | Prob > F = 0.0000 | | |
| | Variable | Coef. | Std. Err. | t | P>t |
| Explanatory variables | V2 | 0.8324 | 0.1671 | 4.98 | 0.000*** |
| | V3 | 0.1037 | 0.0160 | 6.47 | 0.000*** |
| | V6 | 0.059 | 0.0038 | 15.83 | 0.000*** |
| Control variables | V4 | 0.0006 | 0.00009 | 6.43 | 0.000*** |
| | V5 | -0.0015 | 0.0003 | -5.14 | 0.000*** |
| | _cons | -0.5933 | 0.0285 | -20.78 | 0.000*** |

## 6 Conclusion

To further analyze the causes of regional differences, this paper uses the Theil index panel data from 2004 to 2021 to establish a fixed effects model. The results show that both government forces and market forces significantly affect the differentiated development of regional circulation innovation systems. The result of robust regression is also the same, which shows that China's regional circulation innovation system is driven by the two-wheel drive of the government and the market.

The research of this paper shows that in the case of significant differences in circulation innovation and development in different regions, all regions should strengthen the integration of "effective Government" and "efficient market", give full play to the role of government policies and market mechanisms, tap the potential of innovation and transformation, and build a coordinated development pattern based on value symbiosis and complementary advantages. On the one hand, full play is given to the policy-leading role of "effective Government" in technological innovation and transformation of achievements, supportive industrial policies are formulated to promote the circulation power of technological innovation of enterprises, financial support for the transformation of technological innovation is increased, and technological innovation awards, tax incentives, and other means are used to encourage technological innovation personnel and enterprises to innovate. On the other hand, play is given to the resource allocation role of the "efficient market", regional market segmentation is broken, the flow market of technological innovation factors and the circulation market of innovation achievements across the country are coordinated, a "nation-region" multi-level circulation innovation system is established, and the utilization rate of technological innovation resources is improved.

## Acknowledgments

This work was supported by the Guangdong Office of Philosophy and Social Science, Guangzhou Philosophy, and Social Science development "the 14th Five-Year Plan" joint project [GD22YYJ10]; The Guangzhou Office of Philosophy and Social Science, Guangdong Province Philosophy and Social Sciences "the 14th Five-Year Plan" Youth Project [2022GZGJ41].

## References

[1] Chen, J., Yang, Z. & Zhang, Y.Y. (2022). China's Scientific and Technological Innovation under the Perspective of Common Prosperity: Logical Turn and Paradigm Innovation. Reform, (01):1-15. https://chn.oversea.cnki.net/KCMS/detail/detail.aspx?dbcode=CJFD&dbname=CJFDLAST2022&filename=REFO202201001&uniplatform=OVERSEA&v=SB3SkfVwbL6a7pHZs-7bFM8dMftoOLitOlIHSpZheq8m2c4We2pI7eGBFiCVDadX.

[2] Liu, Z. (2022). Analysis of the Internal Requirements for Government to Play a Better Role in Integrated Innovation. Science and Technology Management Research, 42(04):29-35. https://chn.oversea.cnki.net/kcms/detail/detail.aspx?dbcode=CJFD&filename=KJGL202204004&dbname=CJFDLAST2022&uid=WEEvREcwSlJHSldTTEYySnoxK2t4dmw2ajU3bVFTYjVZVDA0Q1g1RWlMMD0=$9A4hF_YAuvQ5obgVAqNKPCYcEjKensW4IQMovwHtwkF4VYPoHbKxJw!!

[3] Chen, J., LI, Z.D. & Zhang, Y.Y. (2023). Theoretical Framework and Long-term Mechanism of Central and Local Government Joint Efforts to Solve "Neck Jamming" Technology Problems from the Perspective of Co-innovation. Journal of Shaanxi Normal University (Philosophy and Social Sciences Edition), 52(01):141-152. DOI: 10.15983/j.cnki.sxss.2023.0113.

[4] Cui, W.J., Sun, C. & Chen, G. (2021). Does Distance Produce Beauty: The Influence of Government-Enterprise Relationship on Enterprise Co-innovation. Science of Science and Management of S.& T., 42(06):81-101. https://chn.oversea.cnki.net/KCMS/detail/detail.aspx?dbcode=CJFD&dbname=CJFDLAST2021&filename=KXXG202106006&uniplatform=OVERSEA&v=l7GP6UXRIU1FChzZ6Q4kAk-zSTK9iH0HXlSUHXuXLoIJ4j91UXH060ZLXtA-_wWG.

[5] Chen, J. & Yang, Z. (2021). Breakthroughs in key and core technologies from the perspective of co-innovation: theoretical framework and implementation path. Journal of Social Sciences, (05):58-69. DOI: 10.13644/j.cnki.cn31-1112.2021.05.005.

[6] Lu, X.X. & Li, L. (2021). Strengthen the dominant position of enterprise innovation and enhance the technological innovation ability of enterprises. Study and Practice, (03):30-44. DOI: 10.19624/j.cnki.cn42-1005/c.2021.03.004.

[7] Garrigos, J.A. & Molina, M.D. (2020). Integrating Customers and Suppliers in Retail Co-innovation Mercadona, the Spanish grocery retailer, uses a co-innovation approach in which customers and suppliers collaborate to create new products. Research-technology management, 05(63):33-41. DOI: 10.1080/08956308.2020.1733887.

[8] Klimas, P. & Czakon, W. (2020). Gaming innovation ecosystem: actors, roles, and co-innovation processes. Review of Managerial Science, 03(16):2213-2259. DOI: 10.1007/s11846-022-00518-8.

[9] Yu, F., Chen, J., Tong, L., Yang, L.S., Ji, Y.H., Wan, P. & Cheng, L. (2020). Study on Technology Development Trend and Industrial Chain Integration Innovation Countermeasures of Ceramic Sheet in China Based on Patent Analysis. China Ceramics, 56(11):39-43. DOI: 10.16521/j.cnki.issn.1001-9642.2020.11.006.

[10] Abhari, K., Davidson, E.J. & Xiao, B. (2017). Co-innovation platform affordances Developing a conceptual model and measurement instrument, Industrial Management & Data Systems, 05(117): 873-895. DOI: 10.1108/IMDS-05-2016-0156.

[11] Coutts, J., White, T., Blackett, P., Rijswijk, K., Bewsell, D., Park, N., Turner, J.A. & Botha, N. (2017). Evaluating a space for co-innovation: Practical application of nine principles for co-innovation in five innovation projects. Outlook On Agriculture, 02(46): 99-107. DOI: 10.1177/0030727017708453.

[12] Chen, C.P. & Li, Y. (2020). Integrating IP protection and Collaborative Innovation from the Incomplete Contractual Route. Journal of Soochow University (Philosophy & Social Science Edition), 41(06):66-73+199. DOI: 10.19563/j.cnki.sdzs.2020.06.009.

[13] Fang, Y.Y., Liu, J.J. & Cao, R.N. (2020). Evolving Mechanism of Enterprises Accommodating Innovation in Open Collaborative Innovation System-Based on Multiple Knowledge Sharing Evolutionary Game Model. Inquiry into Economic Issues, (09):171-180.https://chn.oversea.cnki.net/kcms/detail/detail.aspx?dbcode=CJFD&filename=JJWS202009017&dbname=CJFDLAST2020&uid=WEEvREcwSlJHSldTTEYySnoxK2t4dmw2ajU3bVFTYjVZVDA0Q1g1RWlMMD0=$9A4hF_YAuvQ5obgVAqNKPCYcEjKensW4IQMovwHtwkF4VYPoHbKxJw!!

[14] Ombrosi, N., Casprini, E. & Piccaluga, A. (2019). Designing and managing co-innovation: the case of Loccioni and Pfizer. European Journal of Innovation Management, 04(22):600-616. DOI: 10.1108/EJIM-09-2018-0196.

# Analysis of the functional regression problem of temperature on Ozone concentration

**Yanqing Gou and Xuemin Zi***

School of Science, Tianjin University of Technology and Education, Tianjin, 300074 China

*Corresponding author's e-mail: zi_xuemin@aliyun.com

**Abstract:** With the continuous improvement of computer storage capabilities and online observation technology, functional data emerged as the times required. Among the various research methods, regression analysis of functional data stands out as one of the most commonly employed techniques. This study employed Fourier spline basis functions to smooth the gathered data and preprocess it to establish a functional linear regression model with scalar response. Specifically, we investigated temperature regression on Ozone concentration across 17 cities in Henan Province in 2022. In our functional regression prediction, we compared the results obtained with and without the addition of a roughness penalty. Through this comparison, we observed that the absence of a roughness penalty led to overfitting, while including the penalty yielded more detailed results. Moreover, the coefficient of determination $R^2$ approached one more closely when using the roughness penalty, indicating a better-fitting regression effect. Lastly, for those interested in monitoring the regression residuals of the estimated Ozone concentration, we provide a detailed flowchart outlining the process for monitoring these residuals.

## 1. Introduction

Over the past few years, due to rapid economic development and accelerated urbanization, Ozone has emerged as a significant secondary pollutant impacting urban air quality, following PM2.5. Excessive ground-level Ozone concentration poses serious threats to human health, the ecological environment, and the growth and yield of crops. Research indicates that Ozone concentration is closely correlated with meteorological factors. Among these factors, temperature notably influences changes in Ozone concentration [1]. Therefore, this paper employs daily average temperature to estimate and predict Ozone concentration by establishing a functional regression model.

## 2. Functional regression model

The majority of work in functional predictor regression relies on the functional linear model (FLM):

$$y_i = \beta_0 + \int x_i(t)\beta(t)dt + \varepsilon_i \tag{1}$$

where $y_i, i = 1,...,N$ is the continuous response, $\beta_0$ is the intercept, $x_i(t)$ is the functional predictor, $\beta(\mathrm{t})$ is the functional coefficient, and $\varepsilon_i \sim N(0,\sigma^2)$ is the residual error [2][3][4].

We fit equation (1) by minimizing the sum of squares with penalties to estimate the coefficients using the roughness penalty.

$$PEN_{\lambda}(\beta_0,\beta)=\sum[y_i-\beta_0-\int x_i(t)\beta(t)dt]^2+\lambda\int[L\beta(t)]^2dt \quad (2)$$

where $\lambda$ is a smoothing parameter, $L\beta(t)$ is a differential operator [4][5][6].

To monitor the residual $e=y-\hat{y}$, we contemplate applying Hotelling's $T^2$ control chart to the coefficients derived from functional principal component decomposition [7]of e, i.e.,

$$e_K=\sum_{i=1}^{K}\xi_i^e\phi_i^e \quad (3)$$

where $K$ denotes the count of retained principal components, the scores $\xi_i^e=\langle e,\phi_i^e\rangle$ and the principal components $[\phi_i^e]$ are the eigenfunctions corresponding to the eigenvalues $[\lambda_i^e]$.

The Hotelling's statistic $T^2$ can be delineated as follows [8]:

$$T_e^2=\xi^e\sum\nolimits_{\xi^e}^{-1}\xi^e \quad (4)$$

where $\sum_{\xi^e}=diag(\lambda_1^e,...,\lambda_K^e)$ is the variance-covariance matrix of $\xi^e=(\xi_1^e,...,\xi_K^e)^T$.

## 3. Application

We collected daily average temperature and Ozone concentration data from 17 different cities in Henan Province from January 1, 2022, to December 31, 2022. We evaluated the impact of daily mean temperature on Ozone concentrations by analyzing these data using a functional linear regression model. By how close the coefficient of determination $R^2$ is to 1, we evaluate the fitting effect of the regression model. These data come from the website http://www.tianqihoubao.com.

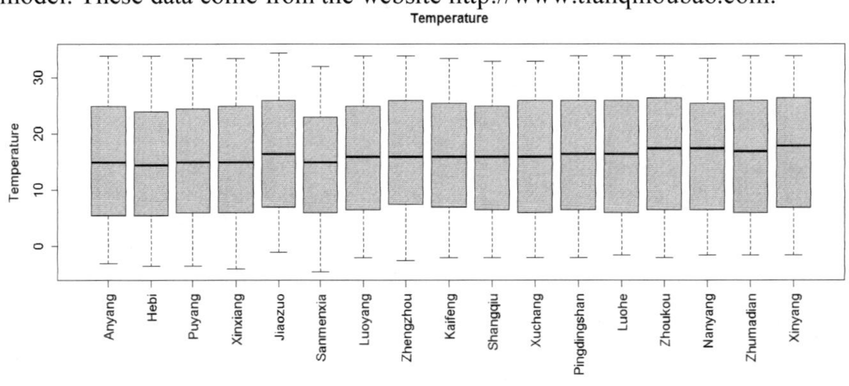


**Figure 1.** Box plot of daily average temperatures in 17 cities in Henan Province in 2022.

Figure 1 shows that the annual average temperature in various cities across Henan Province ranges between 10 and 20 °C. The highest temperatures exceed 30 °C, while the lowest are below 0 °C.

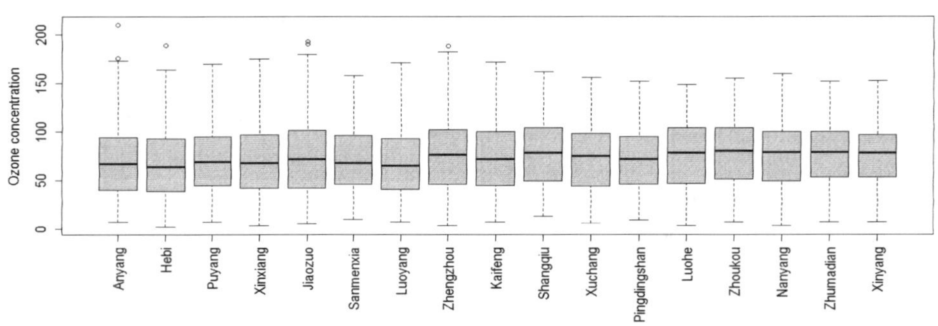


**Figure 2.** Box plot of Ozone concentration in 17 cities in Henan Province in 2022.

As depicted in Figure 2, the annual average concentration of Ozone in various cities across Henan Province in 2022 falls within the range of 50 to 100 μg/m³. However, the highest concentrations are between 150 and 200 μg/m³.

### *3.1 Data smoothing*

We use Fourier spline basis functions to smoothly fit the collected data and draw a functional change curve based on R language programming, as shown in the figure 3 below.

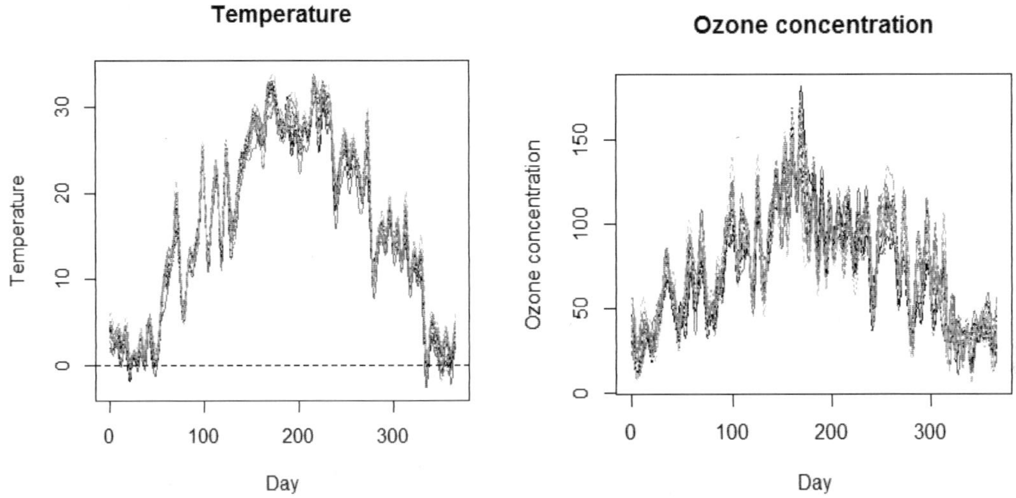


**Figure 3.** Change curves of daily average temperature and Ozone concentration in 17 cities in Henan Province in 2022.

As depicted in Figure 3, a smooth function curve is generated upon applying the Fourier spline basis function to the original data. The visualization illustrates the seasonal and periodic variation characteristics of both temperature and Ozone levels. Typically, temperatures and Ozone concentrations are higher during summer compared to the other three seasons. Conversely, winter exhibits the opposite trend, while spring and autumn fall into a moderate range.

### *3.2 Comparison of two estimation methods*

This paper investigates a functional linear regression model with a scalar response. Initially, the functional Ozone concentration data requires preprocessing. We averaged the Ozone concentration data from 17 cities in Henan Province for 2022, resulting in the annual average Ozone concentration for these 17 cities in Henan Province for 2022. We used the result as the scalar response of the function linear regression model. According to the established model and the data, we can correspond them, $y_i$ corresponds to the Ozone concentration, and $x_i$ corresponds to the daily average temperature. We use the fRegress function in the R language fda package to perform functional regression analysis on Ozone concentration [4].

We extracted the coefficient $\beta(t)$ estimated by the functional regression model from the running results, i.e., $\hat{\beta}(t)$, and used R language to draw the image of the function $\hat{\beta}(t)$. Figure 4 estimate $\beta(t)$ s the mean Ozone concentration based on temperature using 10 Fourier basis functions. Figure 5 estimate $\beta(t)$ s the mean Ozone concentration based on temperature prediction using the roughness penalty method with a smoothing parameter $\lambda\ 10^5$.

Journal of Physics: Conference Series **2813** (2024) 012019 doi:10.1088/1742-6596/2813/1/012019

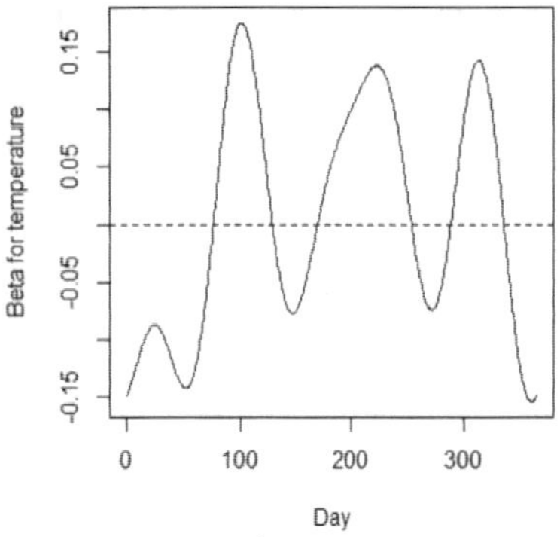


**Figure 4.** Estimate of coefficient $\beta(t)$.

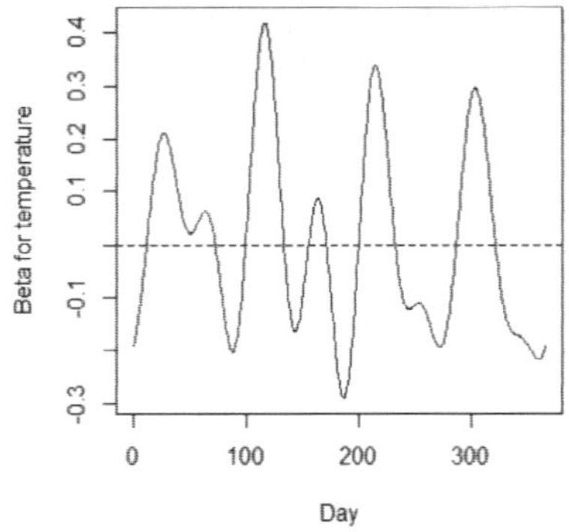


**Figure 5.** Estimation of coefficient $\beta(t)$ using roughness penalty.

The comparison between Figures 4 and 5 shows that the coefficient estimation $\beta(t)$ becomes more nuanced after incorporating the roughness penalty. Figure 5 exhibits more pronounced peaks, valleys, and fluctuations than Figure 4. For instance, around the 200th day, representing the summer season, Figure 4 indicates a positive coefficient $\beta$, suggesting a positive relationship between temperature and Ozone concentration, with the coefficient still close to its peak. Conversely, Figure 5 depicts $\beta$ dipping below zero during this period, indicating a negative correlation between temperature and Ozone concentration and nearing a valley. These two states are nearly opposite. However, since summer often sees substantial rainfall, with climate factors such as precipitation and sunshine impacting Ozone concentration, incorporating a roughness penalty ensures more accurate estimation results. Failure to do so might lead to excessive smoothing.

We extracted the predicted average Ozone concentration based on air temperature and mapped it on a graph in R language with the original collected data. Figure 6 shows the observed average Ozone concentration and the linear regression predicted value of the average Ozone concentration according to temperature; then, we can calculate the value of the determination coefficient $R^2$ as 0.869. Figure 7 shows the observed average Ozone concentration corresponding to the linear regression prediction value of the average Ozone concentration based on temperature using the roughness penalty method. The calculated $R^2$ value is 0.995. The calculated $R^2$ values are all greater than 0.5, and we know the established regression model is effective.

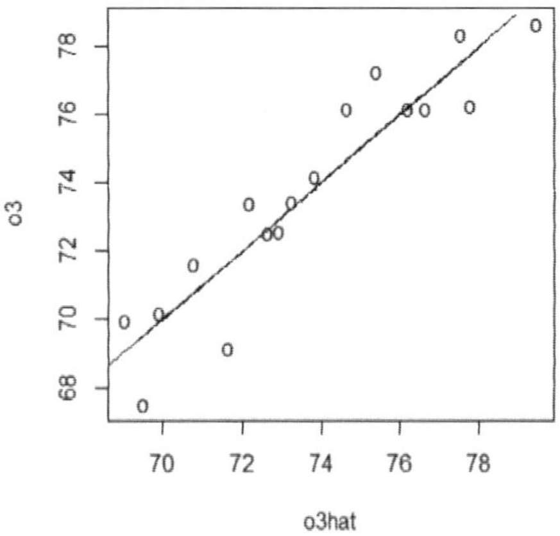


**Figure 6.** Correspondence between true value and predicted value.

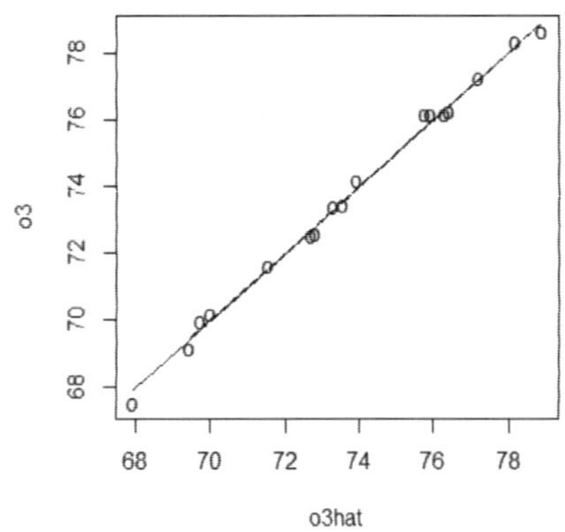


**Figure 7.** Correspondence of true values to predicted values estimated with roughness enalty.

Judging from the proximity of the calculated $R^2$ to 1 and the corresponding outcomes depicted in Figures 6 and 7, it's evident that the estimation results obtained using the roughness penalty method are more precise, yielding a superior fitting effect.

*3.3 Residual monitoring.*

To promptly detect anomalies in Ozone pollution and mitigate Ozone pollution, we monitor its residuals and establish control limits. The process steps are shown in Figure 8 [8]:

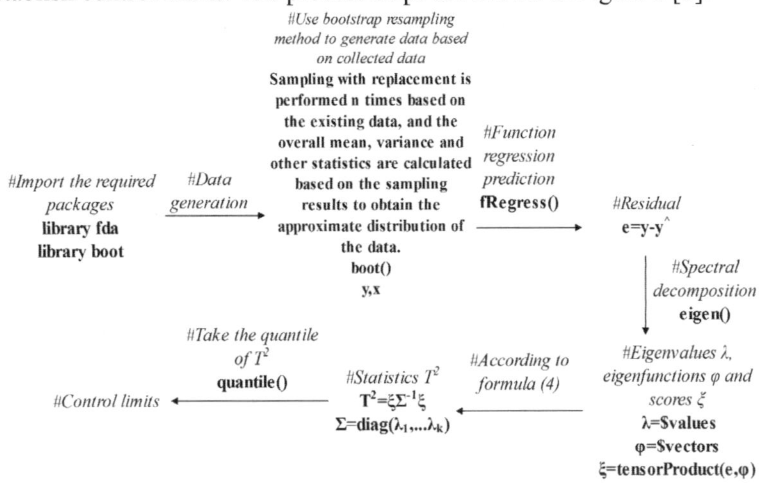


**Figure 8.** Residual monitoring flow chart implemented in R language.

## 4. Conclusion

This paper employs a functional linear regression model to conduct a regression analysis on the Ozone concentration in Henan Province, in which the Ozone concentration is regressed and predicted based on the daily average temperature. It further compares the results with and without including the roughness penalty method. When the roughness penalty method is not added, the coefficient of determination $R^2$ is 0.869, and the regression result demonstrates relative accuracy and proximity to the true value. Upon incorporating the roughness penalty method, the $R^2$ value elevates to 0.995, approaching unity, thus suggesting an improved fitting effect. All values $R^2$ exceeded 0.5, and the final findings indicate the viability of using temperature as a predictor for Ozone concentration regression. Upon comparing the fitting graph and the proximity of $R^2$ 1, we find that utilizing the roughness penalty method leads to a better fitting effect and more accurate results. However, the acknowledgment that temperature impacts Ozone concentration and the potential health hazards posed by excessive Ozone levels underscore the importance of monitoring both Ozone concentration and the regression residuals associated with it—a subject warranting further investigation.

## Acknowledgments

The authors express their gratitude to all reviewers for their valuable feedback and suggestions, which significantly enhanced the quality of this paper. This research was supported by NNSF (11771332).

## References

[1] Li X Y, Li S J and Liu P F 2018 Spatial and temporal variations of Ozone concentrations in China in 2016. *ACTA SCIEN. CIRCUM,* 38(4): 1263-1274.

[2] Morris J S 2015 Functional regression. *ANNU REV STAT APPL*, 2: 321-359.

[3] Ramsay J O and Silverman B W 2005 *Functional data analysis, 2nd ed.* Springer, New York. pp 260-277

[4] Ramsay J O, Hooker G and Graves S 2009 *Functional data analysis with R and MATLAB*. Springer, New York. pp 131-146

[5] Ramsay J O and Silverman B W 1997 *Functional data analysis*. Springer, New York. pp 157-177

[6] Cardot H, Ferraty F, and Sarda P 2003 Spline estimators for the functional linear model. *STAT SINICA*, 13: 571-591.

[7] Hsing T and Eubank R 2015 *Theoretical foundations of functional data analysis, with an introduction to linear operators*. John Wiley & Sons. pp 116-190

[8] Centofanti F, Lepore A, Menafoglio A, Palumbo B and Vantini S 2020 Functional regression control chart. *TECHNOMETRICS*, 63(3): 281-294.

# Some properties of effect algebra

**Wei Li*, Wen Han, Yiyue Sun, Jiangtao Li and Jiaying Wang**

Foundation Department, Engineering University of PAP, Xi'an 710086, China

*Corresponding author's e-mail: liweidezuiaihao@163.com

**Abstract:** As a kind of algebraic structure, effect algebra is the foundation of quantum structure and quantum measurement. Sequential effect algebra is a special type of effect algebra that plays an important role in the fundamental research of Quantum Computation. Therefore, we study the related theory of effect algebra in this paper. Finally, four properties and special examples of effect algebra and sequence effect algebra are presented, as well as three properties of effect algebra level sum.

## 1. Introduction

Quantum logic is an important branch of quantum mechanics. As a kind of algebraic structure, effect algebra plays an important role in the basic research of quantum structure and quantum measurement, and has been widely concerned. An effect algebra is a partially ordered algebraic structure that can be seen as an extension of quantum logic or orthogonal modular lattices. To better understand the properties of effect algebra, we have established generalized effect algebra, weak effect algebra, sequence effect algebra, expression of effect algebra, congruence, and ideal of effect algebra based on effect algebra [1-5]. Effect algebra has attracted more and more attention from mathematicians, logicians, and computer experts, and its content is also in the process of continuous improvement [6-10].

Although the research on effect algebra is becoming increasingly sophisticated, there are still many problems. This article provides some properties and special examples of effect algebra and sequence effect algebra and discusses some properties of effect algebra level sum.

**Definition 1.1** [1] $G$ is a set with two special elements $0,1$( $0 \neq 1$ ), and $\oplus$ is a partial binary operation on $G$, then $(G,\oplus,0,1)$ is an algebra of effects if it satisfies the following conditions :

(1) If $d_{11} \oplus f_{22}$ is defined, then $d_{11} \oplus f_{22}$ is defined and $d_{11} \oplus f_{22} = f_{22} \oplus d_{11}$.

(2) If $d_{11} \oplus f_{22}$ and $(d_{11} \oplus f_{22}) \oplus k_1$ are defined, then $f_{22} \oplus k_1$ and $d_{12} \oplus (f_{22} \oplus k_1)$ are defined and $(d_{11} \oplus f_{22}) \oplus k_1 = d_{11} \oplus (f_{22} \oplus k_1)$.

(3) For every $d_1 \in G$ there is a unique $f_1 \in G$ such that $d_1 \oplus f_1$ exists and $d_1 \oplus f_1 = 1$.

(4) If $f_1 \oplus 1$ is defined, thus $f_1 = 1$.

$(G,\oplus,\circ,0,1)$ is an effect algebra, if $a_1 \oplus a_2$ is defined, we denote it as $a_1 \perp a_2$. For every $a_1 \in E$, there is a unique $b_1 \in E$ in Condition (3), which makes $a_1 \oplus b_1 = 1$, we denote it as $a_1' = b_1$.

If $a_1, a_2 \in G$, and exists $c_1 \in G$, such that $a_1 \perp c_1$ and $a_1 \oplus c_1 = a_2$, denote $a_1 \leq a_2$ and $c = a_2 \Theta a_1$, then $\leq$ is a partial order relation of $(G,\oplus,0,1)$, and $a_{11} \perp a_{22}$ if and only if $a_{11} \leq a_{22}'$. $(G,\oplus,0,1)$ is a lattice about $\leq$, thus $(G,\oplus,0,1)$ is a lattice effect algebra. $(G,\oplus,0,1)$ is an ordered set about $\leq$, thus it is called ordered effect algebra.

$(G,\oplus,\circ,0,1)$ is an effect algebra, $d\in E$, if $d\wedge d'=0$, thus $d$ becomes the exact element of $E$. The set $G_s=\{d\in G: d\wedge d'=0\}$ is the set of all exact elements in $G$.

**Definition 1.2** [2] $(G,\oplus,\circ,0,1)$ is an effect algebra. $\circ$ is one of its binary operations, satisfying the following conditions:

(1) For every $a_{11}\in E$, if $d_{11}\perp f_{11}$, thus $a_{11}\circ d_{11}\perp a_{11}\circ f_{11}$ and

$$a_{11}\circ(d_{11}\oplus f_{11})=a_{11}\circ d_{11}\oplus a_{11}\circ f_{11}. \quad (1)$$

(2) For every $d_1\in E$, $1\circ d_1=d_1$.

(3) If $d_{11}\circ f_{11}=0$, then $d_{11}\circ f_{11}=f_{11}\circ d_{11}$.

(4) For every $c_1\in G$, if $d_{11}\circ f_{11}=f_{11}\circ d_{11}$, thus $f_{11}\circ d'_{11}=d'_{11}\circ f_{11}$ and

$$d_{11}\circ(f_{11}\circ c_{11})=(d_{11}\circ f_{11})\circ c_{11}. \quad (2)$$

(5) If $c_{11}\circ a_{11}=a_{11}\circ c_{11}, c_{11}\circ d_{11}=d_{11}\circ c_{11}$, thus $c_{11}\circ(a_{11}\circ d_{11})=(a_{11}\circ d_{11})\circ c_{11}$. When $a_{11}\perp d_{11}$, there is $c_{11}\circ(a_{11}\oplus d_{11})=(a_{11}\oplus d_{11})\circ c_{11}$, then $(G,\oplus,\circ,0,1)$ is called a sequential effect algebra.

$(G,\oplus,\circ,0,1)$ is sequential effect algebra, if $d\circ f=f\circ d$, we said $d,f$ to be sequence-independent, recorded as $d\mid f$.

**Lemma 1.1** [5] $(G,\oplus,\circ,0,1)$ is sequential effect algebra, $a_{11},b_{11},d_{11}\in G$, and the following proposition holds:

(1) If $a_{11}\perp b_{11}$, $a_{11}\perp d_{11}$, and $a_{11}\oplus d_{11}=a_{11}\oplus d_{11}$, thus $b_{11}=d_{11}$.

(2) $d_{11}\in G_s \Leftrightarrow d_{11}\circ d_{11}=d_{11} \Leftrightarrow d_{11}\circ d'_{11}=0$.

(3) If $d_1\in G_s$, thus $a_1\le d_1$ if and only if $a_1=a_1\circ d_1=d_1\circ a_1$.

## 2. Conclusions about sequential effect algebra

**Theorem 2.1** The sequential effect algebra does not satisfy the elimination law, i.e. when $a\circ c=b\circ c$, $a=b$ cannot be obtained.

Example 2.1 Let the set $C_4=\{0,a_{11},b_{22},1\}$. If $a_{11}\oplus b_{22}=1$, then $(C_4,\oplus,\circ,0,1)$ is a sequential effect algebra, so that $G_s=C_4$. From $1\circ b_{22}=b_{22}\circ b_{22}=b_{22}$, $1\ne b_{22}$ is deduced, so the sequential effect algebra does not satisfy the elimination law.

**Theorem 2.2** $(G,\oplus,\circ,0,1)$ is a sequential effect algebra. If $d_1\circ k_1=k_1\circ d_1$, and $d_1\vee k_1<f_1\in E_s$,t thus $d_1=k_1$.

Proof Because $d_1\vee k_1<f_1\in E_s$, thus $d_1=d_1\circ f_1=f_1\circ d_1=k_1$.

**Theorem 2.3** $(G,\oplus,\circ,0,1)$ is a sequential effect algebra. If $G_s$ is commutable, thus $G_s$ is the subsequence effect algebra of $G$.

Proof (1) $d\in G_s$, $d'\in G_s$,i.e. $E_s$ is closed concerning operations.

(2) $\forall d_{11},f_{22}\in E_s$, and $d_{11}\mid f_{22}$, then $d_{11}\circ d_{11}=d_{11}$, $f_{22}\circ f_{22}=f_{22}$. If $d_{11}\perp f_{22}$, i.e. $d_{11}\le f_{22}'$, thus

$$d_{11}\circ f_{22}\le f_{22}'\circ f_{22}=0,$$

So

$$d_{11}\circ f_{22}=f_{22}\circ d_{11}=0.$$

$$\begin{aligned}(d_{11}\oplus f_{22})\circ(d_{11}\oplus f_{22})&=((d_{11}\oplus f_{22})\circ d_{11})\oplus((d_{11}\oplus f_{22})\circ f_{22})\\&=(d_{11}\circ(d_{11}\oplus f_{22}))\oplus(f_{22}\circ(d_{11}\oplus f_{22}))\\&=(d_{11}\circ d_{11})\oplus(d_{11}\circ f_{22})\oplus(f_{22}\circ d_{11})\oplus(d_{11}\circ f_{22})=d_{11}\oplus f_{22},\end{aligned}$$

i.e. $G_s$ is closed concerning $\oplus$ operations.

(3) For every $d_{11}, f_{22} \in G_s$, and $d_{11} | f_{22}$,

$$(d_{11} \circ f_{22}) \circ (d_{11} \circ f_{22}) = d_{11} \circ (f_{22} \circ (d_{11} \circ f_{22})) = d_{11} \circ ((f_{22} \circ f_{22}) \circ d_{11})$$
$$= d_{11} \circ (f_{22} \circ d_{11}) = (d_{11} \circ d_{11}) \circ f_{22} = d_{11} \circ f_{22},$$

i.e. $G_s$ is closed concerning $\circ$ operations.

So $G_s$ is the sub-sequence effect algebra of $G$.

**Theorem 2.4** $(G, \oplus, 0, 1)$ is effect algebras. If $a_{11} \in E_s, \forall b_{22} \le a_{11}$, then $b_{22}$ doesn't have to belong to $G_s$.

Example 2.2

$$G = \{0, k_{11}, k_{11} \oplus k_{11}, q_{22}, q_{22} \oplus q_{22}, k_{11} \oplus q_{22}, k_{11}', q_{22}', (k_{11} \oplus q_{22})', (q_{22} \oplus q_{22})', (k_{11} \oplus q_{22})', 1\}$$
$$G_s = \{0, \mathrm{k}_{11} \oplus k_{11}, q_{22} \oplus q_{22}, k_{1'1} \oplus q_{22}, (k_{11} \oplus q_{22})', (k_{11} \oplus q_{22})', (k_{11} \oplus q_{22})', 1\},$$

$k_{11} \oplus k_{11} \in G_s$ and $k_{11} \le k_{11} \oplus k_{11}$, but $k_{11} \notin G_s$.

**Definition 2.1** [5] $(G, \oplus, \circ, 0, 1)$ is the sequential effect algebra, $K \subseteq G$. If $0, 1 \in K$, and $(K, \oplus, \circ, 0, 1)$ is a sequential effect algebra, then define $K$ is a sub-sequence effect algebra of $(G, \oplus, \circ, 0, 1)$.

The subset $K$ of $(G, \oplus, \circ, 0, 1)$ is a subsequence effect algebra if and only if the operation is closed concerning three partial binary operations$'$, $\circ$, $\oplus$.

If $\Phi \ne K \subseteq G$, thus there is a minimal sub-sequence effect algebra $\overline{K}$ in $G$ and $K \subseteq \overline{K}$, we said $\overline{K}$ is a sequential effect algebra generated with $K$.

Example 2.3 $(G, \oplus, \circ, 0, 1)$ is a sequential effect algebra, $A = \{f_{11}\} \subseteq E$, find $\overline{A}$.

Solution (1) If $f_{11} \in G_s$, and $f_{11}$ is not an isotropic element. $A_1 = \{f_{11}, f_{11}'\}$,

$$A_2 = \{f_{11}, f_{11}', 0, 1\}, A_3 = A_4 = \cdots = A_2,$$

Then $\overline{A} = A_2 = \{f_{11}, f_{11}', 0, 1\}$.

(2) If $a_1 \in G_s$, and $f_{11} \perp f_{11}$.

$$A_1 = \{f_{11}, f_{11}', 2f_{11}\}, A_2 = \{f_{11}, f_{11}', 2f_{11}, (2f_{11})', 0, 1\}, A_3 = A_4 = \cdots = A_2,$$

then $\overline{A} = A_2 = \{f_{11}, f_{11}', 2f_{11}, (2f_{11})', 0, 1\}$.

**Theorem 2.5** The two sub-effect algebras with intersection are sub-effect algebra.

Proof $(G, \oplus, 0, 1)$ is an effect algebra, $A, B$ is the two sub-effect algebras of $G$, i.e.. $K \le G, S \le G$.

(1) $0 \in K \cap S$, $1 \in K \cap S$;

(2) If $\forall f \in K \cap S$, thus $f' \in K \cap S$;

For $f \in K$, then $f' \in K$, and for $f \in S$, then $f' \in S$. So $f' \in K \cap S$;

(3) Easy to prove, $\oplus$ satisfies the associative law.

(4) Easy to prove, $\oplus$ satisfies commutative law.

(5) $\oplus$ satisfies the closure: if $d_1, k_2 \in K \cap S$, and $d_1$ and $k_2$ can be added, then

$$d_1 \in K, k_2 \in K, d_1 \oplus k_2 \in K; d_1 \in S, k_2 \in S, d_1 \oplus k_2 \in S.$$

So $d_1 \oplus k_2 \in K \cap S$.

**Theorem 2.6** The union of two sub-effect algebras of an effect algebra is not necessarily an effect algebra.

Proof $(D, \oplus, 0, 1)$ is an effect algebra, $G, H$ is the two sub-effect algebras of $D$, ie $G \le D, H \le D$, where

$$D = \{0, d_1, d_1', p_1, p_1', c_1, c_1', 1\},$$
$$d_1 \oplus d_1' = p_1 \oplus p_1' = c_1 \oplus c_1' = 1,$$

$$d_1 \oplus p_1 = c_1,\ d_1 \oplus c_1' = p_1',\ p_1 \oplus c_1' = d_1',$$
$$G = \{0, d_1, d_1', 1\},\ H = \{0, p_1, p_1', 1\},$$
$$G \cup H = \{0, d_1, d_1', p_1, p_1', 1\}.$$

If $d_1 \perp p_1$, then $d_1 \oplus p_1 = c_1 \notin G \cup H$. $\oplus$ operation is not closed.

$$E_S = \{0, d_1, d_1', p_1, p_1', c_1, c_1', 1\} = E.$$

## 3. Level sum of sequential effect algebra

**Theorem 3.1** [3] $\{(E_i, \oplus_i, 0_i, 1_i)\}_{i \in I}$ is a family of effect algebras, and

$$E_i' \cap E_j' = \Phi (i \neq j).$$

Taking $0,1 \notin E_i'$, for every $i \in I$, we define $E = \underset{i \in I}{\cup} E_i' \cup \{0,1\}$. In the set $E$, a partial binary operation is defined as follows:

(1) For every $f \in E_i', f \oplus 0 = 0 \oplus f = f$,

(2) For every $f \in E_i', f \oplus f' = 1$,

(3) For every $d, f \in E_i'$, if $d \perp f$, $b \neq a'$, define $d \oplus f = f \oplus_i d$,

(4) $1 \oplus 0 = 0 \oplus 1 = 1$.

Then $(E, \oplus, \circ, 0, 1)$ to be a sequential effect algebra.

**Definition 3.2** [3] $\{(G_i, \oplus_i, 0_i, 1_i)\}_{i \in I}$ be a family of effect algebras, if $G_i' = G_i \setminus \{0_i, 1_i\}$, if $G_i' \cap G_j' = \Phi (i \neq j)$,

then it is said to be summable, and the effect algebra $(G, \oplus, 0, 1)$ in Theorem 3.1 is called the level sum of $\{(G_i, \oplus_i, 0_i, 1_i)\}_{i \in I}$, denoted $G = HS(G_i, i \in I)$. When $I = \{1, 2, \cdots, n\}$, the level sum of the compatible family $\{G_1, G_2, \cdots, G_n\}$ denote as $HS(G_1, G_2, \cdots, G_n)$.

**Theorem 3.2** $(G, \oplus, \circ, 0, 1)$ is a sequential effect algebra.

$\hat{G} = \{0,1\} \cup \{\hat{g}_1 : g_1 \in G \setminus \{0,1\}\}$,

where $0' = 1$, $1' = 0$, for $\forall g_1 \in G \setminus \{0,1\}$ there is

$$\hat{g}_1' = \hat{g_1'},\ 0 \oplus \hat{g}_1 = \hat{g}_1 \oplus 0 = \hat{g}_1,\ 1 \oplus 0 = 0 \oplus 1 = 1,$$

for every $g_1, k_1 \in G \setminus \{0,1\}$, $g_1 \perp k_1$ there is

$$\hat{g}_1 \hat{\oplus} \hat{k}_1 = \begin{cases} g_1 \oplus k_1 & g_1 \oplus k_1 \neq 1 \\ 1 & g_1 \oplus 1 = 1 \end{cases} \tag{3}$$

thus $(\hat{G}, \hat{\oplus}, 0, 1)$ is an effect algebra, $G \cap \hat{G} = \{0,1\}$. $\mathrm{HS}(G, \hat{G}) = G \cup \hat{G}$ is denoted as the level sum of $G$ and $\hat{G}$.

We define the mapping $\phi$ from $K$ to $\hat{K}$: $K \to \hat{K}$

$$0_k \to 0_{\hat{k}}$$
$$1_k \to 1_{\hat{k}}$$
$$x_1 \to \hat{x}_1 \qquad \forall x_1 \in K \setminus \{0,1\}$$

Define the mapping $\varphi$ from $\hat{K}$ to $K$: $\hat{K} \to K$

$$0_{\hat{k}} \to 0_k$$
$$1_{\hat{k}} \to 1_k$$

$$\hat{x}_1 \to x_1$$

i.e. $\phi\varphi = I_k$, $\varphi\phi = I_{\hat{K}}$.

**Conclusion 3.1** We define $\hat{\circ}$ in $\hat{G}$, $\hat{\circ}$ satisfies

$$0\,\hat{\circ}\,\hat{x}_1 = \hat{x}_1\,\hat{\circ}\,0 = 0, 0\,\hat{\circ}\,1 = 1\,\hat{\circ}\,0 = 0,$$

then $(\hat{G},\hat{\oplus},\hat{\circ},0,1)$ is a sequence effect algebra. If

$$1\,\hat{\circ}\,\hat{x}_{11} = \hat{x}_{11}\,\hat{\circ}\,1 = 0, \quad \hat{x}_{11}\,\hat{\circ}\,\hat{y}_{11} = \begin{cases} x_{11} \circ y_{11} & x_{11} \circ y_{11} \neq 0 \\ 0 & x_{11} \circ y_{11} = 0 \end{cases} \tag{4}$$

then $(\hat{G},\hat{\oplus},\hat{\circ},0,1)$ is the sequential effect algebra.

**Conclusion 3.2** In HS$(G,\hat{G}) = G \cup \hat{G}$, for every $u_{11}, k_{22} \in G \cup \hat{G}$, when $u_{11}, k_{22} \in G$, $u_{11} \perp k_{22}$, there is $u_{11} \overline{\overline{\oplus}} k_{22} = u_{11} \oplus k_{22}$; When $u_{11}, k_{22} \in \hat{G}$, $u_{11} \perp k_{22}$, there is $u_{11} \overline{\overline{\oplus}} k_{22} = u_{11} \hat{\oplus} k_{22}$, and the rest cannot be added. then (HS$(G,\hat{G})$, $\overline{\overline{\oplus}}$, $\overline{\overline{\circ}}$, 0, 1) is a sequential effect algebra.

Proof because $x_{11} \circ \hat{y}_{11} = x_{11} \circ y_{11}$, $\hat{y}_{11} \circ x_{11} = \hat{y}_{11}\,\hat{\circ}\,\hat{x}_{11} = y_{11} \circ x_{11}$, $x_{11} \circ \hat{y}_{11} = \hat{y}_{11} \circ x_{11}$, so

$$x_{11} \circ y_{11} = y_{11} \circ x_{11} = 0.$$

$$u_{11} \overline{\overline{\circ}} k_{22} = \begin{cases} u_{11} \circ k_{22} & u_{11}, k_{22} \in G \\ u_{11}\,\hat{\circ}\,k_{22} & u_{11}, k_{22} \in \hat{G} \\ u_{11} \circ \phi(k_{22}) & u_{11} \in G, k_{22} \in \hat{G} \\ u_{11}\,\hat{\circ}\,\varphi(k_{22}) & u_{11} \in \hat{G}, k_{22} \in G \end{cases},$$

and HS$(G,\hat{G})$ are commutative if and only if for every $k, p \in G$, if $k \circ p = 0$, then $k \mid p$. So (HS$(G,\hat{G})$, $\overline{\overline{\oplus}}$, $\overline{\overline{\circ}}$, 0, 1) is sequential effect algebra.

Example 3.1 If $C_3 = \{0, a, 1\}$, then $\hat{C}_3 = \{0, \hat{a}, 1\}$, the level sum is

$$HS(C_3, \hat{C}_3) = \{0, a, \hat{a}, 1\} = C_4(a, \hat{a}),$$

Where $2a = 2\hat{a} = 1$.

Example 3.2 If $C_4 = \{0, a_1, b_2, 1\}$, then $\hat{C}_4 = \{0, \hat{a}_1, \hat{b}_2, 1\}$, the level sum is

$$\text{HS}(C_4, \hat{C}_4) = \{0, a_1, \hat{a}_1, b_2, \hat{b}_2, 1\},$$

where $2a = 2b = 1$.

Example 3.3 The level sum of $C_3 = \{0, a, 1\}$ and $[0,1]$ is HS$(C_3, [0,1]) = \{a\} \cup [0,1]$.

**Conclusion 3.3** The level sum of two lattice effect algebras is a lattice effect algebra.

Proof If $G, K$ are lattice effect algebras, then for every $u_1, k_2 \in$ HS$(G, K)$,

$$u_1 \vee k_2 = \begin{cases} u_1 \vee_A k_2 & u_1, k_2 \in G \\ u_1 \vee_B k_2 & u_1, k_2 \in K \\ 1 & u_1 \in G, k_2 \in K \text{ or } u_1 \in K, k_2 \in G \end{cases},$$

and $u_1 \vee k_2$ exist, so HS$(G, K)$ is a lattice.

**References**

[1] Gudder S. Greechie R. Uniqueness and order in sequential effect algebras [J]. International Journal of Theoretical Physics, 2005, 44 (7) 755-770. https://doi.org/:10.1007/s10773-005-7054-y.

Journal of Physics: Conference Series **2813** (2024) 012020 doi:10.1088/1742-6596/2813/1/012020

[2] ZHANG Jian-cheng, WANG Guo-jun. Some properties of morphisms in effect algebras [J]. Journal of Mathematical Research & Exposition, 2009, 29(2): 219-226.https://doi.org/:10.3770/j. ISSN: 1000-341X.2009.02.004.
[3] Hájerk P. Metamathematics of fuzzy logic[M]. Dordrecht: Kluwer Academic Publisher, Dordrecht.
[4] Gudder S. Greechie R. Sequential products on effect algebras[J]. Math. Phys, 2002, 49(1):87-111. https://doi.org/: 10.1016/S0034-4877(02)80007-6.
[5] Huaixin Cao, Zhengli Chen, Zhihua Guo. Algebraic Representation of Effects[J]. Science in China: Mathematical Sciences, 2011, 41 (3): 279-286. https://doi.org/: 10.1360/012010-767.
[6] Qiaowei Zhang. Research on Teaching Reform of Multivariate Statistical Analysis Curriculum Based on the Concept of Three Subjects Education[J]. Journal of Yulin University., 2021.31(2) 38-41. https://doi.org/: 10.16752/j. CNKI. jylu.2021.02.009.
[7] Ivan Chajda, Radomír Halaš, Helmut Länger. The logic induced by effect algebras [J]. Soft Computing, 2020 24:14275–14286. https://doi.org/10.1007/s00500-020-05188-w.
[8] Yali Wu, Xia Li, Jing Wang. Notes on Sharp and Principal Elements in Effect Algebras[J]. Order. 2023.Published online:1-12. https://doi.org/10.1007/s11083-022-09619-1.
[9] Jenčová A., Pulmannová S. Spectrality in Convex Sequential Effect Algebras [J]. International Journal of Theoretical Physics, 2023, 62 (8):193. https://doi.org/:10.1007/s10773-023-05431-8.
[10] Wudong Li, Ying Zhang, Kan He. A Sub-sequential Effect Algchra from the Quantum Programming Theory [J]. Acta Mathematica Sinica Chinese series, 2020, 63(6): 647-654. https://doi.org/: 10.27352/d.cnki.gylgu.2020.001172.

Journal of Physics: Conference Series **2813** (2024) 012021 doi:10.1088/1742-6596/2813/1/012021

# Evading model ownership resolution through reconstructive neuron pruning

**Jun Wu***

Software Engineering Institute, East China Normal University, Shanghai, 200062, China

*Corresponding author's e-mail: 51215902063@stu.ecnu.edu.cn

**Abstract.** Due to training a deep learning model requires a lot of resources, we regard the trained model as the intellectual property (IP) of its trainer. However, researchers have shown that deep learning models are susceptible to model-stealing attacks. Recently, a method to detect theft by embedding external features has been proposed. We argue that this method is similar to backdooring the model and still has deficiencies in protecting model IP. In this paper, we present a method that removes external features to evade such detection. Initially, we get clean data by employing model inversion and adaptive selection. Subsequently, we utilize this data to exclude neurons associated with external features, while simultaneously guaranteeing that the model's accuracy remains relatively unchanged. The empirical data demonstrates that our method can help different types of adversaries evade ownership verification of external features. In addition, our method can further enhance the attack effect under watermark-based defense.

## 1. Introduction

Deep learning models have achieved great success in many tasks [1,2], such as image classification, face recognition, speech recognition, autonomous driving, natural language processing, etc. However, training a deep learning model with good performance requires a large amount of computing resources and manually labeled samples. This process of training model is very expensive, so the trained model can be considered the intellectual property of its trainer. Researchers have recently discovered that it is possible to train a replica of a model, even without knowledge of the model parameters or training data. Besides, [3] have shown that the attacker can obtain a copy that is closer, sometimes better than the victim model. This has seriously damaged the intellectual property of the model owners, which needs to be taken seriously.

To prevent the model stealing, there are three defense strategies. Firstly, we add appropriate perturbations to the output. It not only increases the attacker's stealing cost [5] but also affects the accuracy of legitimate users. Secondly, we verify the model fingerprint [6-8]. Model fingerprint mainly develops some inherent model attributes that are unique characteristics of the model, and defenders use these characteristics to detect theft. Thirdly, we inject model watermarking [9,10]. To prevent the model from being stolen from the target model, watermarking is intended to introduce specific information into the model training process that differs from the original training data.

Recently, one way to safeguard the intellectual property of models has been proposed by [11]. They proposed a method named embedding external features. They incorporated the style-transferred images into the original dataset for model training, then trained a meta-classifier through the benign model and the victim model, and finally used hypothesis testing for ownership verification.

This paper presents a method to help the adversary evade ownership verification of external features. We performed adversarial neuron perturbation on the adversary model and found that both the adversary model and the backdoor model have similar behaviors. Inspired by this, we believed that the adversary model contained a backdoor which is the external features. Therefore, we can prune neurons related to external features to evade ownership verification of external features. However, we had no information about external features. [12] proposed Reconstructive Neuron Pruning (RNP), which can solve the above problem. RNP requires a part of clean training data. In general, we could get training data from model inversion, but the training data was messy, which didn't meet the requirement of RNP. To improve training data quality, we proposed a distribution score to filter out high-quality data. More details will be shown in Section 3. We have two main contributions as follows:

- The proposal of method. Our method can help all adversaries evade ownership verification of external features. The effectiveness of the method also shows that external features are similar to backdoor features.
- Experiments results. In addition to external features, we also set up two watermark-based defense methods. Experimental results show that our method may have a certain generality for this defense method. We believe that with the emergence of more powerful pruning technology, watermark-based defense methods will face greater challenges.

## 2. Related work

### *2.1. Model stealing*

Model stealing aims to obtain a replica of the deployed model to illicitly obtain the victim's intellectual property. Generally speaking, there are three types of model stealing techniques depending on how much access the adversary is given. The following are these categories:

- Dataset Attacks ($\mathcal{A}_{\mathcal{D}}$): The adversary can acquire the dataset of the target model. An alternative model can be trained using knowledge distillation [13].
- Model Attacks ($\mathcal{A}_{\mathcal{M}}$): The adversary possesses unrestricted access to the victim model, including the gradients, hyper-parameters, and structure. One option is to utilize data-free knowledge distillation [14] to train a different model, which is based on zero-shot learning. Another option is to fine-tune the target model using their dataset.
- Query Attacks ($\mathcal{A}_{\mathcal{Q}}$): Only the target model is queryable by the attacker. Based on various feedbacks, it can be further subdivided into label-query [4] and logit-query [3]. In the context of label-query, the adversary can assign labels to unlabeled data using the target model, and subsequently use this labeled data to train its model. In the context of logit-query, the adversary can reduce the distance between the output of the alternative model and the target model to train alternative models.

### *2.2. Embedding external features*

Embedding external features is a verification-based defense method [11]. It consists of three stages, constructing the dataset, training the meta-classifier, and ownership verification.

*2.2.1. Training set.* Specifically, The benign training set is represented as $\mathcal{D} = \{(x_i, y_i)\}_{i=1}^{N}$, where each sample consists of an input image $x_i$ and its corresponding label $y_i$. The style image $x_s$ and the trained style transformer $T: \mathcal{X} \times \mathcal{X} \rightarrow \mathcal{X}$ are also introduced. To generate the transformed version of the selected samples, denoted as $\mathcal{D}_t = \{(x', y) \mid x' = T(x, x_s)\}$, the defender randomly chooses $\gamma\%$ of the samples ($\mathcal{D}_s$) from the benign training set $\mathcal{D}$. The transformation process involves applying the trained style transformer T to the input images x. We make $\mathcal{D}_b = \mathcal{D} \backslash \mathcal{D}_s$ training set be $\mathcal{D}_b \cup \mathcal{D}_t$, and the defenders use it to train victim models.

*2.2.2. Meta-classifier.* They use the model weight gradients as the meta-classifier's input when training it. The training set $\mathcal{D}_s$ of meta-classifier shows as follows:

$$\mathcal{D}_c = \{(g_v(x'), +1) \mid (x', y) \in \mathcal{D}_t\} \cup \{(g_b(x'), +1) \mid (x', y) \in \mathcal{D}_t\}. \quad (1)$$

where $g_v(\cdot)$ is the sign of the victim model weight gradient and $g_b(\cdot)$ is the sign of the benign model weight gradient. The meta-classifier $clf(\cdot)$ is trained using the following equation:

$$\min_{w} \sum_{(s,t)\in\mathcal{D}_c} \mathcal{L}(clf(\boldsymbol{s}), t) \quad (2)$$

*2.2.3. Ownership verification.* In their ownership verification approach, the researchers utilize hypothesis testing. Specifically, they randomly select $m$ distinct transformed images from the dataset $\mathcal{D}_t$ and perform pair-wise T-tests to calculate the corresponding p-values. The null hypothesis is rejected when the computed p-value is below the predetermined significance level α. Besides, a confidence score $\Delta\mu$ is computed to indicate the level of confidence in the verification process. A higher $\Delta\mu$ value corresponds to a greater degree of confidence in the verification outcome. For details about $\Delta\mu$, we can refer to [11].

## 3. Method

*3.1. Overview*

[11] conducted an experiment in which they partitioned a small portion of private data from the training data as a carrier for external features. They performed style transfer by the private data to embed the external features and then trained a meta-classifier to determine whether a model was stolen from a victim model. They argued that the model that was taken from the victim ought to be aware of the features that the victim's model had picked up, i.e., the adversary model $M_A$ should contain the knowledge of external features, and then compare the external features using the meta-classifier. We believe that this is utilizing model overfitting, similar to the backdoor model. Embedding external features is similar to injecting a backdoor in some ways. To validate our idea, we considered private data as backdoor data and performed adversarial neuron perturbation [15] by the adversary model $M_A$ as follows:

$$f_{M_A}(x; (1+\boldsymbol{\delta}) \odot \boldsymbol{w}, (1+\boldsymbol{\xi}) \odot \boldsymbol{b}) \quad (3)$$

where $f_{M_A}(\cdot)$ is the output of the perturbed model, and $\odot$ is neuron-wise multiplication. Each perturbation component of $\boldsymbol{\delta}$ and $\boldsymbol{\xi}$ acts on the same neuron weight $\boldsymbol{w}$ and bias $\boldsymbol{b}$, respectively. Formally,

$$(1+\boldsymbol{\delta}) \odot \boldsymbol{w} = \left[\left(1+\delta_1^{(1)}\right) w_1^{(1)}, \cdots, \left(1+\delta_{n_L}^{(L)}\right) w_{n_L}^{(L)}\right] \quad (4)$$

$$(1+\boldsymbol{\xi}) \odot \boldsymbol{b} = \left[\left(1+\xi_1^{(1)}\right) b_1^{(1)}, \cdots, \left(1+\xi_{n_L}^{(L)}\right) b_{n_L}^{(L)}\right] \quad (5)$$

The notations $w_k^{(l)}$ and $b_k^{(l)}$ represent the k-th neuron in the l-th layer. $\delta_k^{(l)}$ and $\xi_k^{(l)}$ are the corresponding perturbations. The impact of adversarial neuron perturbations on sensitivity is closely associated with the injected backdoor [15]. We can use this property to determine whether the adversary model has backdoor behavior or not. The left side of Figure 1 shows that as the perturbation increases, the error rate of the adversary model also increases, which exhibits the same behavior as the BadNets. They are associated with various forms of potential backdoors. Further, we explored the distribution of each class and randomly selected 10% from the training data of the cifar10 airplane class to do style transfer, and the airplane is our target class. Then we embedded the external features to get the External Features (EF) model and fine-tuned the EF model to get the adversary model $M_A$. The right side of Figure 1 shows that $M_A$ tends to misclassify the input sample $X$ as an airplane under the adversarial neuron perturbations. This is consistent with the research conducted by [15].

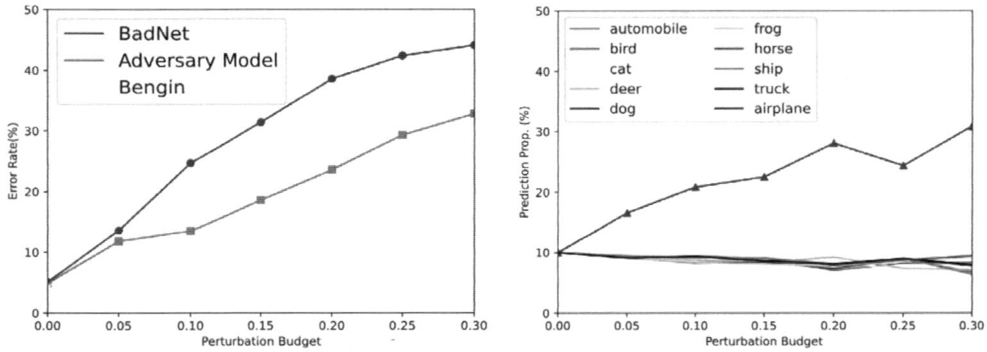


**Figure 1.** The picture on the left is the error rate of different models under neuron perturbations. The picture on the right is the proportion of different classes in predictions by an adversary model under neuron perturbations. The dataset is CIFAR10.

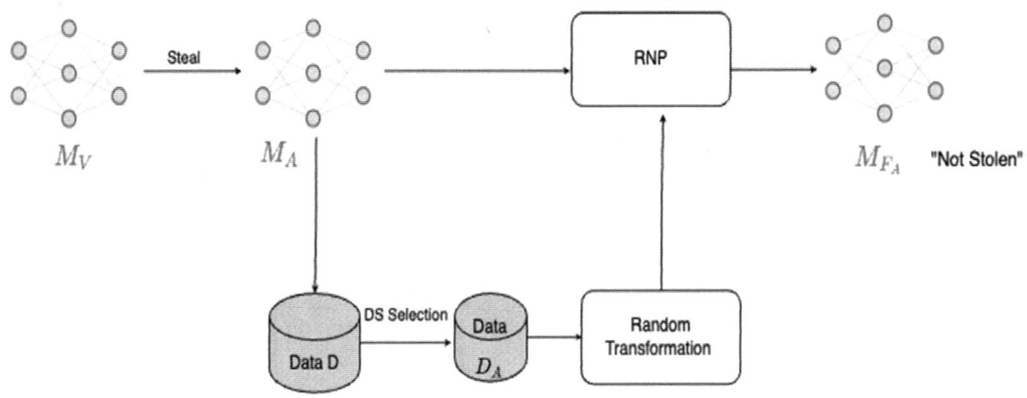


**Figure 2.** The main flow of our method.

Our method can be divided into two main steps. Firstly, we inverse the adversary model to obtain a set of data. We propose a distribution score (DS) metric to select data $D_A$ within the distribution, which will serve as the attack data for the adversary. Before using $D_A$, a set of random transformations is required to mitigate potential external features that may be translated. Secondly, we utilize this data to perform Reconstructive Neuron Pruning (RNP) on the adversary model $M_A$. Then we obtain a model $M_{F_A}$ capable of evading defender detection.

### *3.2. Model inversion and distribution score (DS)*

The Reconstructive Neuron Pruning (RNP) requires a portion of clean data from the training set, but the adversary does not have any information about the training set. we can invert the data from the model. But in our setting, the adversary can not use $M_V$ to invert the data. Since the adversary model is a replica of the victim model, we can invert the data using the adversary model $M_A$ rather than the victim model $M_V$. Our model inversion attack is based on deep inversion. An arbitrary target label y and a random noise $\hat{x}$ are supplied, and the image is synthesized by optimizing:

$$\min_{\hat{x}} \alpha_l \mathcal{L}(\hat{x}, y) + \alpha_p R_{prior}(\hat{x}) + \alpha_f R_{feature}(\hat{x}) \tag{6}$$

where $\alpha_l$, $\alpha_p$, and $\alpha_f$ are the coefficients of the terms, $R(\cdot)$ is an image regularization term, $\mathcal{L}(\cdot)$ is cross-entropy loss. $R_{prior}(\hat{x})$ is used to penalize the total variance and $l_2$ norm of $\hat{x}$. At the same time, to increase the images' fidelity and bring them as near to the original training set as feasible, $R_{feature}(\hat{x})$ is used to effectively enhance feature similarity at all levels. After we generate several images, we need to select the images closest to the original training set (i.e. in distribution). To achieve this, we propose a distribution score, as follows:

$$DS(x) = \frac{p_{top1}(x) - \frac{1}{m}}{var(\boldsymbol{p}_{-}(x)) - \frac{1}{m} + 1} \tag{7}$$

where $m$ is the number of classes, and $p_{top1}(x)$ represents the largest component in the prediction vector. $\boldsymbol{p}_{-}(x)$ represents the variance of the probabilities of other classes except $p_{top1}(x)$, the other items ensure that $0 \leq DS(x) \leq 1$. As can be seen, the probability distribution turns into a one-hot vector when the $DS(x)$ reaches its maximum value, and it becomes uniform when it reaches its smallest value. We believe that the larger the $DS(x)$ of the data is, the closer the distribution of the original dataset is. Although this is only a rough estimate, it is sufficient for our problem.

### *3.3. Random transformation*

We generate images from the adversary model, but by adversarial neuron perturbation, we can know that the adversary model has potential backdoor feature knowledge. If the generated image contains such features, it does not meet our requirements for clean data in RNP and has an impact on the results. The impact will be discussed in the next section, so we need to mitigate the impact of these potential backdoor features. Actually, we can achieve this by just randomly transforming $D_A$. A series of fundamental transformations is compounded into a random transformation. $\{T(\cdot; \theta_i)\}_{i=1}^{n}$, i.e., $T(\cdot; \theta) = T_n(T_{n-1}(\cdots T_1(\cdot; \theta_1); \theta_{n-1}); \theta_n)$. $\theta_i$ is the corresponding transformation coefficient of $T_i$.

### *3.4. Reconstructive neuron pruning (RNP)*

RNP consists of three main components: Unlearning, Recovering, and Pruning. Our main idea is that we should remove neurons that are more sensitive to external features. These neurons can be considered as backdoor neurons. One way is to maximize the loss of the defender's private dataset to remove external features. However, the private dataset is not available to the adversary. We can only expose neurons related to external features from the other side. In RNP, we do not need a defender private dataset. The fundamental aspect of RNP involves a reconstructive learning procedure. Neuron Unlearning (NU) is the first step in this procedure, which involves unlearning the neurons in the defense data. Filter Recovering (FR) is then used to recover those neurons on the same data [12]. RNP has an asymmetric learning and recovery process, exposing more neurons related to external features and performing better purification in the pruning step.

*3.4.1. Unlearning.* The defender's training set consists of two parts $\mathcal{D} = \mathcal{D}_c \cup \mathcal{D}_t$, with $\mathcal{D}_c = \{(x_c^{(i)}, y_c^{(i)})\}_{i=1}^{N}$ representing the clean data and $\mathcal{D}_t = \{(x_t^{(j)}, y_t^{(j)})\}_{j=1}^{T}$ representing the data after style transfer. We use part of the $\mathcal{D}_c$ as defense data $D_d$, $\mathcal{D}_c$ can be obtained through model inversion and DS selection, i.e., $D_A$ can replace $\mathcal{D}_c$. Unlike the process of training a model, a model can be unlearned for a specific task by adjusting its loss function based on the task-specific data [12].

*3.4.2. Recovering.* Restoring the clean features (features of clean samples $\mathcal{D}_c$) that were eliminated in the preceding stage is the aim of this step. Efficient recovery can be achieved by updating the unlearned model to reduce its classification loss on defense data. Within the framework of RNP, the recovery process directs its attention toward filters rather than individual neurons. To identify potential backdoor filters and their corresponding neurons, a filter mask is employed [12].

*3.4.3. Pruning.* This step mainly uses the mask obtained in the previous step to screen neurons for pruning. As part of the recovery process, the filter mask's element values are initially set to one and then adjusted to ensure they are within the range of [0, 1] by clipping them. After recovering, low values of the filter mask close to zero indicate that the filter (and its associated neurons) contains a large proportion of neurons that may be associated with external features. This point can be used as the basis for pruning [12].

## 4. Experiments

Our experiments rely on the source code of the research proposed by [11]. Therefore, the settings in our experiments are consistent with the research conducted by [11] as follows.

### *4.1. Setup*

*4.1.1. Model selection.* For our research, we used the WRN-28-10 architecture as the victim model for the CIFAR-10. In addition, we employed ResNet as the victim model for the ImageNet. Regarding the adversary model, we followed the methodologies outlined in the work by [8] to train the adversary model using various techniques.

*4.1.2. Defense setup.* In addition to the external features, we also experiment with two other methods based on backdoor verification, model watermarking with BadNets and Entangled Watermarks. We feed the adversary models obtained through the five different methods to our system, to obtain a model $M_{F_A}$. We then conduct ownership verification on the model $M_{F_A}$, observing that the p-value.

*4.1.3. Model Inversion.* The number of images we need is 1% of the original training set. Specifically, about CIFAR10, we generated 5000 images and then filtered the top 500 images as $D_A$ according to the DS. Similarly, with ImageNet12, we generate 2500 images and take the top 125 images as $D_A$.

*4.1.4. Evaluation metric.*Our evaluation metrics are aligned with the research conducted by [11]: $\Delta\mu$ and p-value. The level of confidence in the verification increases with a larger $\Delta\mu$. We randomly select 10 distinct transformed images from the dataset $D_t$ for conducting pair-wise T-tests and computing the corresponding p-values. Defenders can assert that the adversary model has been illicitly obtained from the victim model when the calculated p-value falls below the predetermined significance level $\alpha$ (e.g., 0.05 or 0.01).

### *4.2. Results*

We make ownership verification with hypothesis tests in different five attack methods. Table 1 displays the values of $\Delta\mu$ and p-value, which indicate the level of confidence in the hypothesis test. If the p-value is less than the specified significance level $\alpha$, the adversary model will be classified as stolen. Besides, for all defense methods we provide the victim (called the source) and an independent model trained with other datasets as a reference for the effectiveness of the defense method.

**Table 1.** Results of $\Delta\mu$ and p-value using 10 samples on five adversary models.

**(a) Results on CIFAR10.**

| Model Stealing | | BadNets | | Entangled Watermarks | | External Features | |
|---|---|---|---|---|---|---|---|
| | | $\Delta\mu$ | p-value | $\Delta\mu$ | p-value | $\Delta\mu$ | p-value |
| Victim | Source | 0.91 | $10^{-12}$ | 0.99 | $10^{-35}$ | 0.98 | $10^{-37}$ |
| $\mathcal{A}_\mathcal{D}$ | Distillation | $-10^{-3}$ | 0.32 | 0.01 | 0.33 | 0.52 | $10^{-6}$ |
| | Distillation+ours | $-10^{-2}$ | 0.33↑ | $10^{-4}$ | 0.48↑ | $10^{-4}$ | 0.23↑ |
| $\mathcal{A}_\mathcal{M}$ | Zero-shot | $10^{-25}$ | 0.22 | $10^{-3}$ | $10^{-3}$ | 0.46 | $10^{-5}$ |
| | Zero-shot+ours | $10^{-27}$ | 0.25↑ | 0.05 | 0.29↑ | $10^{-3}$ | 0.16↑ |
| $\mathcal{A}_\mathcal{Q}$ | Fine-tuning | $10^{-25}$ | 0.28 | 0.35 | 0.01 | 0.46 | $10^{-10}$ |
| | Fine-tuning+ours | $10^{-25}$ | 0.29↑ | $10^{-5}$ | 0.51↑ | $10^{-5}$ | 0.37↑ |
| | Label-query | $10^{-27}$ | 0.20 | $10^{-5}$ | 0.62 | 0.44 | $10^{-5}$ |
| | Label-query+ours | $10^{-28}$ | 0.20 | $10^{-8}$ | 0.66↑ | $10^{-3}$ | 0.18↑ |
| | Logit-query | $10^{-27}$ | 0.23 | $10^{-6}$ | 0.64 | 0.42 | $10^{-5}$ |
| | Logit-query+ours | $10^{-29}$ | 0.27↑ | $10^{-7}$ | 0.65↑ | $10^{-6}$ | 0.41↑ |
| Benign | Independent | $10^{-20}$ | 0.33 | $10^{-22}$ | 0.68 | $10^{-9}$ | 1.00 |

**(b) Results on ImageNet12.**

| Model Stealing | | BadNets | | Entangled Watermarks | | External Features | |
|---|---|---|---|---|---|---|---|
| | | $\Delta\mu$ | p-value | $\Delta\mu$ | p-value | $\Delta\mu$ | p-value |
| Victim | Source | 0.97 | $10^{-20}$ | 0.99 | $10^{-25}$ | 0.97 | $10^{-33}$ |
| $\mathcal{A}_D$ | Distillation | $10^{-8}$ | 0.22 | $10^{-6}$ | 0.20 | 0.43 | $10^{-5}$ |
| | Distillation+ours | $10^{-8}$ | 0.26↑ | $10^{-7}$ | 0.37↑ | $10^{-4}$ | 0.26↑ |
| $\mathcal{A}_M$ | Zero-shot | $10^{-9}$ | 0.33 | $10^{-7}$ | 0.46 | 0.40 | $10^{-7}$ |
| | Zero-shot+ours | $10^{-10}$ | 0.35↑ | $10^{-8}$ | 0.58↑ | $10^{-4}$ | 0.31↑ |
| $\mathcal{A}_Q$ | Fine-tuning | $10^{-7}$ | 0.20 | 0.28 | 0.01 | 0.44 | $10^{-4}$ |
| | Fine-tuning+ours | $10^{-8}$ | 0.27↑ | 0.01 | 0.18↑ | $10^{-5}$ | 0.56↑ |
| | Label-query | $10^{-10}$ | 0.39 | $10^{-7}$ | 0.41 | 0.43 | $10^{-4}$ |
| | Label-query+ours | $10^{-12}$ | 0.46↑ | $10^{-8}$ | 0.59↑ | $10^{-3}$ | 0.25↑ |
| | Logit-query | $10^{-9}$ | 0.34 | $10^{-6}$ | 0.38 | 0.49 | $10^{-5}$ |
| | Logit-query+ours | $10^{-11}$ | 0.41↑ | $10^{-7}$ | 0.45↑ | $10^{-3}$ | 0.21↑ |
| Benign | Independent | $10^{-27}$ | 0.49 | $10^{-30}$ | 0.73 | $-10^{-5}$ | 1.00 |

As shown in Table 1, the p-values of almost all adversary models using our method have become larger. We find that in an adversary model without using our method, BadNets cannot effectively detect theft. This may be because, during the theft process, the triggers used for verification are wiped out, resulting in a reduction in the number of neurons associated with the backdoor features [11]. After using our method, there is no big change in the p-value of BadNets. This may be because the adversary model no longer contains explicit backdoor features, making it difficult to locate backdoor neurons. This is intuitive. Besides, under defenses Entangled Watermarks and External Features, the p-values of the adversary model using our method exceed the significance level $\alpha$ (0.01). We successfully evade ownership verification. Table 2 shows that the accuracy of the adversary model before and after RNP on the testing set will not drop significantly. Although the adversary loses some test accuracy, she or he obtains a model that can evade ownership verification.

**Table 2.** The accuracy of the victim model and five adversary models. The $ACC_b$ and $ACCa$ mean before and after RNP.

| | CIFAR10 | | ImageNet12 | |
|---|---|---|---|---|
| | $ACC_b$ | $ACCa$ | $ACC_b$ | $ACCa$ |
| Victim | 95.5 | - | 90.8 | - |
| Zero-shot | 88.1 | 85.5 | 78.6 | 75.8 |
| Distillation | 90.6 | 86.1 | 82.3 | 79.7 |
| Fine-tune | 95.6 | 93.2 | 90.1 | 87.3 |
| Label-query | 85.2 | 83.8 | 80.5 | 77.5 |
| Logit-query | 91.9 | 89.8 | 86.8 | 85.2 |

### 4.3. *Effect of random transformation*

The defense method is an external feature and the data set is CIFAR10. We compare the p-values with and without Random Transformation (RT). Table 3 shows that without RT, the p-value decreases. Zero-shot, Label-query, and Logit-query cannot evade ownership detection. This is because $D_A$ contains external features. This does not meet the definition of clean data. RNP treats the backdoor features as clean features for unlearning and recovering, resulting in the inability to accurately locate the corresponding neurons. RT can just suppress the expression of potential non-clean features.

**Table 3.** The results with and without random transformation.

| | External Features | |
|---|---|---|
| | p(+RT) | p(-RT) |
| Distillation+ours | 0.23 | 0.02 |
| Zero-shot+ours | 0.16 | $10^{-3}$ |
| Fine-tuning+ours | 0.37 | 0.11 |
| Label-query+ours | 0.18 | $10^{-6}$ |
| Logit-query+ours | 0.41 | $10^{-4}$ |

## 5. Conclusion

This study presents a novel approach to circumvent the process of verifying ownership. To get the approximate clean training data, firstly, we employ an adversary model to produce a dataset. Then, we employ an adaptive approach to choose synthetic images that closely resemble the distribution of the original training set. Finally, we use RNP to prune neurons related to non-clean features. These non-clean features are related to ownership verification. Experiments show that our method can help different adversaries evade external feature detection. Exploring better external features is valuable work.

## References

[1] Torfi, A., Shirvani, R. A., Keneshloo, Y., Tavaf, N., and Fox, E. A. (2020). Natural language processing advancements by deep learning: A survey. arXiv preprint arXiv:2003.01200. https://doi.org/10.48550/arXiv.2003.01200

[2] Maini, P., Yaghini, M., and Papernot, N. (2021). Dataset inference: Ownership resolution in machine learning. arXiv preprint arXiv:2104.10706. https://doi.org/10.48550/arXiv.2104.10706.

[3] Orekondy, T., Schiele, B., and Fritz, M. (2019). Knock-off nets: Stealing functionality of black-box models. In Proceedings of the IEEE/CVF conference on computer vision and pattern recognition, pages 4954-4963. http://openaccess.thecvf.com/content_CVPR_2019/html/Orekondy_Knockoff_Nets_Stealing_Functionality_of_Black-Box_Models_CVPR_2019_paper.html.

[4] Jagielski, M., Carlini, N., Berthelot, D., Kurakin, A., and Papernot, N. (2020). High accuracy and high fidelity extraction of neural networks. In the 29th USENIX Security Symposium (USENIX Security 20). Boston .pages 1345–1362. https://www.usenix.org/conference/usenixsecurity20/presentation/jagielski.

[5] Lee, T., Edwards, B., Molloy, I., and Su, D. (2019). Defending against neural network model stealing attacks using deceptive perturbations. In 2019 IEEE Security and Privacy Workshops (SPW). San Francisco, CA, USA.pages 43–49. 10.1109/SPW.2019.00020.

[6] Cao, X., Jia, J., and Gong, N. Z. (2021). Ipguard: Protecting the intellectual property of deep neural networks via fingerprinting the classification boundary. In Proceedings of the 2021 ACM Asia Conference on Computer and Communications Security. Melbourne. pages 14–25. https://doi.org/10.1145/3433210.3437526.

[7] Lukas, N., Zhang, Y., and Kerschbaum, F. (2019). Deep neural network fingerprinting by conferrable adversarial examples. arXiv preprint arXiv:1912.00888. https://doi.org/10.48550/arXiv.1912.00888.

[8] Maini, P., Yaghini, M., and Papernot, N. (2021). Dataset inference: Ownership resolution in machine learning. arXiv preprint arXiv:2104.10706. https://doi.org/10.48550/arXiv.2104.10706.

[9] Guo, J. and Potkonjak, M. (2018). Watermarking deep neural networks for embedded systems. In 2018 IEEE/ACM International Conference on Computer-Aided Design (ICCAD). San Diego .pages 1–8. 10.1145/3240765.3240862.

[10] Darvish Rouhani, B., Chen, H., and Koushanfar, F. (2019). Deepsigns: An end-to-end watermarking framework for ownership protection of deep neural networks. In Proceedings of the Twenty-Fourth International Conference on Architectural Support for Programming Languages and Operating Systems. pages 485–497.https://doi.org/10.1145/3297858.3304051.

[11] Li, Y., Zhu, L., Jia, X., Jiang, Y., Xia, S.-T., and Cao, X. (2022). Defending against model stealing via verifying embedded external features. In Proceedings of the AAAI Conference on Artificial Intelligence, volume 36. Washington, DC, USA. pages 1464–1472. https://doi.org/10.1609/aaai.v36i2.20036.

[12] Li, Y., Lyu, X., Ma, X., Koren, N., Lyu, L., Li, B., and Jiang, Y.-G. (2023). Reconstructive neuron pruning for backdoor defense. arXiv preprint arXiv:2305.14876. https://doi.org/10.48550/arXiv.2305.14876.

[13] Hinton, G., Vinyals, O., and Dean, J. (2015). Distilling the knowledge in a neural network. arXiv preprint arXiv:1503.02531. https://doi.org/10.48550/arXiv.1503.02531.

[14] Micaelli, P. and Storkey, A. J. (2019). Zero-shot knowledge transfer via adversarial belief matching. Advances in Neural Information Processing Systems, 32. https://proceedings.neurips.cc/paper_files/paper/2019/hash/0266e33d3f546cb5436a10798e657d97-Abstract.html.

[15] Wu, D. and Wang, Y. (2021). Adversarial neuron pruning purifies backdoored deep models. Advances in Neural Information Processing Systems, 34: 16913–16925. https://proceedings.neurips.cc/paper/2021/hash/8cbe9ce23f42628c98f80fa0fac8b19a-Abstract.html.

# A comprehensive analysis of evaluating robustness and generalization ability of models in AES

**Ruijie Liu[1], Xiaoyi Wang[2], Jie Liu[2, 3*] and Jianshe Zhou[2]**

[1] Information Engineering College, Capital Normal University, Beijing, China
[2] Language Intelligence Research Center, Capital Normal University, Beijing, China
[3]School of Information Science and Technology, North China University of Technology, Beijing, China
Ruijie Liu and Xiaoyi Wang, These authors contributed equally to this work

[*]Corresponding author's e-mail: liujxxxy@126.com

**Abstract:** Large language models are capable of impressive performance in standard benchmark testing, but are often fragile when deployed in real-world scenarios and cannot be deployed in many scenarios directly, such as navigation and autonomous driving. When some adversarial samples require continuous learning of new data distributions, the robustness and generalization abilities of the model require detailed evaluations and further mining. Therefore, it is very necessary to use tools to evaluate the robustness and generalization ability of the model before deploying models in real-world scenarios. The robustness and generalization capabilities of models were extensively assessed before their implementation in Automatic essay Scoring. This was achieved through a comprehensive series of experiments aimed at validating the efficacy of various models. Our findings underscored the significance of different models and their impact, providing valuable insights for practitioners seeking to forecast the performance of deployed models across diverse distributions.

## 1. Introduction

Evaluation benchmarks play a crucial role in artificial intelligence. Traditional natural language processing (NLP) benchmarks primarily access specific and relatively simple abilities. Large language models (LLMs) have demonstrated various new abilities and shifted the evaluation request to more general and complex skills, such as model robustness and generalization abilities. Consequently, to accommodate the new era of LLMs, new benchmarks have been proposed, aiming at exploring the diverse capabilities of LLMs.

Advances in NLP have enabled models to achieve high accuracy with independent and identically distributed (IID) training and test data [1]. However, analysis reveals a critical drawback: these models lack robustness against data corruption and distribution changes, and may rely on spurious patterns for prediction. In practical applications, these vulnerabilities limit the resolution of unseen data and hinder the deployment of reliable systems in real-world scenarios. For example, after the training of the navigation model is completed, if it is directly applied to the actual traffic scenes, the model likely will not learn all possible road conditions. Therefore, the model does not make correct judgments and actions, and the resulting loss is immeasurable.

The primary objective of model evaluation is to enable users to assess the performance of their models accurately for deployment readiness, not to discover that the model performs poorly or even suffers irreparable damage after actual deployment. Based on this, we start from the field of automatic essay scoring, apply evaluation tools, and evaluate the current essay scoring models with good performance and LLMs. The evaluation results are useful for model update and deployment in real scenarios. Practitioners can take the results as a guide to avoid a series of losses caused by poor performance of the model under a new environment (new data distribution).

A central challenge is how to evaluate the model and judge whether it is suitable before putting it into use. Our significant insight is that practitioners can utilize evaluation tools to test the model and then decide whether to use it based on the results. In this way, practitioners can combine the results with their own experience to choose subsequent operations, thus avoiding irreparable mistakes and losses to a large extent.

In summary, the main contributions are as follows:

1. We construct the Chinese Essay Dataset and English essay dataset respectively to evaluate model performance.
2. We test several state-of-the-art AES models and LLMs, and the results demonstrate that the robustness and generalization ability of models still need to be improved, guiding practitioners in using the models.
3. We propose that combining robustness and generalization capabilities before applying the model to real-world scenarios can better evaluate the model's performance and reduce the probability of unrecoverable losses after the model is officially used.

## 2. Related work

### *2.1 Model evaluation*

Model robustness signifies its capacity to remain stable and reliable under different circumstances. Specifically, a robust model can maintain its function and performance in the face of some unexpected or abnormal situations, and will not crash or fail due to these abnormal situations. The model's generalization ability pertains to its capability to perform effectively on unseen data, showcasing its adaptability to novel datasets. Both of them can reflect the reliability of the model in the face of "unwelcome" data. By evaluating the robustness and generalization of the model, practitioners can understand the model's reliability more clearly.

### *2.2 Large language models (LLMs)*

It turns out that the effectiveness of LMs in performing various tasks is largely affected by the size of their architectures. This trend develops to models like GPT-4 [2] with trillions of parameters. As LLMs scale, they can suddenly "unlock" new capabilities, which cover a wide range of tasks.

However, the rapid development of LLMs has also brought some hazards and instability, and they still have certain limitations. Users can use chatbots with coding capabilities to create complex malware with unprecedented ease, while also using coordinated chatbot networks on social media platforms at scale to distort public discourse.

Therefore, the research on the robustness and generalization ability evaluation of LLMs is very urgent. Many researchers have begun to study the robustness and generalization ability evaluation of LLMs, but so far, the number and depth of the research are still not enough.

### *2.3 Automated essay scoring (AES)*

AES is a task employing models to evaluate essays either based on their overall quality or specific characteristics. Amidst the ascent of deep learning, scoring methods from early regression or classification tasks obtain the overall score [3]. Recent advances in deep learning mark a significant stride forward in AES [4], with the added advantage of exploiting automatically learned features from data.

The current performance of AES models is deemed satisfactory, and there remains a scarcity of research addressing the robustness and generalization capabilities of these models. Powers et al. suggest that unless AES systems are explicitly designed to handle adversarial inputs, some writers may subvert them because they understand how the system works [5][6].

## 3. Experiments

### *3.1 Datasets*
We built a Chinese Essay Dataset for Chinese grades 7 to 9, which contains 6, 916 essays from 17 topics and scores in the range of 1-57. Since there are no adversarial samples in the dataset and the whole dataset is IID, we randomly select a part of the essays and shuffle the order of the sentences or delete a sentence randomly to construct a Chinese adversarial sample set for evaluating the robustness of models. We add some generalization data from other tasks into a training set, while the evaluation set remains unchanged to get the training set and evaluation set with different data distributions for evaluating the generalization ability of models. We also use a similar method as dealing with the Chinese Essay Dataset to construct an English adversarial sample dataset and an English essay set with different data distributions.

### *3.2 Models to evaluate*
Our experiments utilize some baseline methods in automated essay scoring during recent years and some LLMs.

On model selection, we select some SOTA AES models and some LLMs:

LC&LSTMT&N: It presents a framework for simultaneously training the coherence model alongside a cutting-edge model, bolstering the credibility of AES approaches against adversarial inputs crafted to deceive the system [7].

HMTS (Hierarchical Multi-task Trait Scorer): It evaluates the quality of writing by modeling traits and can effectively score essays from multiple traits, outperforming other strong baselines [8].

BERT: It is a novel language representation model aimed at pre-training deep bidirectional representations from unlabeled text [9].

ChatGPT: It is a large-scale language model built upon the GPT architecture, which enables a natural conversation flow and can answer various types of questions.

GPT-4: It is a large-scale multimodal model exhibiting human-level performance.

As for evaluation, we also choose some recently proposed evaluation methods:

MANDOLINE: This framework assesses models under distribution shift by employing user-specified slicing functions to adjust estimates [10].

Wang et al. performed a comprehensive assessment of ChatGPT's robustness from both adversarial and out-of-distribution (OOD) viewpoints [11].

To make the results more accurate and convincing, we use Robustness Gym and combine two evaluation metrics to measure the model: accuracy and QWK (Quadrated Weighted Kappa). The accuracy metric gauges the predictive ability of the model for the prompts on the evaluation dataset, while QWK assesses its proficiency in scoring essays. The combination of these two metrics can give a more comprehensive evaluation of the robustness and reliability of the model when facing adversarial examples and data distribution shifts.

## 4. Results
We present the findings from our experiments on both Chinese and English datasets. Table 1 and Table 2 show the impact of data size on model performance under different evaluation methods, where *x means that the amount of data is x times larger than the original amount of dataset. Table 3 and Table 4 respectively represent the use of Robustness Gym to evaluate the model on the Chinese essay dataset and ASAP Dataset.

**Table 1.** The results of using MANDOLINE to evaluate model performance.

| | AVERAGE ESTIMATION ERROR (*0.25) | AVERAGE ESTIMATION ERROR (*0.5) | AVERAGE ESTIMATION ERROR (*0.75) | AVERAGE ESTIMATION ERROR (*1) | AVERAGE ESTIMATION ERROR (*1.5) | AVERAGE ESTIMATION ERROR (*2) |
|---|---|---|---|---|---|---|
| BERT | 17.3% | 16.2% | 13.5% | 7.4% | 6.5% | 6.2% |
| LC&LSTMT&N | 18.9% | 18.4% | 15.9% | 8.5% | 7.9% | 7.4% |
| ChatGPT | 12.3% | 11.5% | 9.4% | 5.8% | 5.6% | 5.2% |
| GPT-4 | 10.7% | 10.2% | 8.6% | 5.2% | 5.0% | 4.9% |

**Table 2.** The results of evaluating model performance.

| | F1 (*0.25) | F1 (*0.5) | F1 (*0.75) | F1 (*1) | F1 (*1.5) | F1 (*2) |
|---|---|---|---|---|---|---|
| BERT | 32.1 | 32.9 | 33.7 | 34.4 | 35.8 | 37.1 |
| LC&LSTMT&N | 30.8 | 31.2 | 32.1 | 33.2 | 34.1 | 36.8 |
| ChatGPT | 48.2 | 50.1 | 51.4 | 53.5 | 55.2 | 56.9 |
| GPT-4 | 56.1 | 57.3 | 58.1 | 58.8 | 59.6 | 60.9 |

**Table 3.** The robustness and generalization performance of models on the ASAP Dataset.

| | Accuracy (%) | | QWK | |
|---|---|---|---|---|
| | rob | gen | rob | gen |
| BERT | 25.2 | 31.4 | 0.524 | 0.546 |
| LC&LSTMT&N | 21.4 | 28.2 | 0.482 | 0.472 |
| ChatGPT | 51.6 | 63.4 | 0.595 | 0.653 |
| GPT-4 | 57.9 | 65.3 | 0.674 | 0.687 |

**Table 4.** The robustness and generalization performance of models on the Chinese Essay Dataset.

| | Accuracy (%) | | QWK | |
|---|---|---|---|---|
| | rob | gen | rob | gen |
| HMTS | 32.5 | 34.2 | 0.531 | 0.462 |
| ChatGPT | 49.8 | 63.7 | 0.624 | 0.659 |
| GPT-4 | 53.4 | 66.1 | 0.696 | 0.679 |

We can see that LLMs outperform other models in robustness and generalization performance. Because LLMs are much larger than other models and they learned a large amount of corpus during training, they are more capable of learning.

The performance of models, that performed well in the AES task, also decays in the experiment. This is because their experimental data are all IID, and they only need to achieve good scoring performance on these IID data. The problem of data distribution shift is not considered, so the model's abilities are not strengthened. Once there are adversarial examples or data distribution shifts, the performance will be greatly tested.

Two LLMs, ChatGPT and GPT-4, achieve the best results, thanks to their large parameter size and good robustness. The results still need to be improved, because the performance of LLMs in a large number of segmentation domains still needs to be strengthened. Once more indistinguishable adversarial examples appear, their generalization ability cannot withstand the test.

At the same time, as the sample size of the data set continues to decrease, the evaluation indicators corresponding to each evaluation method have declined to varying degrees. The smaller the sample size is, the greater the uncertainty of evaluation methods is, making them unreliable and less useful.

By analyzing the results, practitioners can know the performance of models in different environments. If the performance degradation of the model is too obvious, it shows that the model is not very adaptable to the current adversarial samples or different distribution, and practitioners can choose appropriate ways to modify the model to adapt to new challenges.

## 5. Conclusion

We assess the robustness and generalization capabilities of models prior to their deployment in real-world scenarios. A large number of results showed that the advanced AES models have poor robustness and generalization ability, and LLMs have better performance but still need to be improved. We hope that our work can help practitioners determine if the model can be deployed in real-world scenarios and inspire future work on designing AES models that are more robust and have better generalization ability. In the future, we aim to join more specific evaluation metrics according to the characteristics of the task, to evaluate the model ability.

## Acknowledgments

This work is supported by the National Key Research and Development Program of China (2020AAA0109703) and the National Natural Science Foundation of China (62076167, U23B2029).

## References

[1] Moradi M. and Samwald M. Evaluation of the Robustness of Neural Language Models to Input Perturbations [C]// Proceedings of the 2021 Conference on Empirical Methods in Natural Language Processing, 2021: 1, 558-1, 570.

[2] Achiam J., Adler S., Agarwal S., et al. Gpt-4 technical report [J]. arXiv preprint arXiv: 2303.08774, 2023.

[3] Kabra A., Bhatia M., Singla Y. K., et al. Evaluation Toolkit for Robustness Testing of Automatic Essay Scoring Systems [C]// 5th Joint International Conference on Data Science & Management of Data (9th ACM IKDD CODS and 27th COMAD), 2022: 90-99.

[4] Chen Y. and Li X. PMAES: Prompt-mapping contrastive learning for cross-prompt automated Essay scoring [C]// Proceedings of the 61st Annual Meeting of the Association for Computational Linguistics (Volume 1: Long Papers), 2023: 1, 489-1, 503.

[5] Powers D. E., Burstein J. C., Chodorow M., et al. Stumping e-rater: challenging the validity of automated Essay scoring [J]. Computers in Human Behavior, 2002, 18 (2): 103-134.

[6] Yannakoudakis H. and Briscoe T. Modeling coherence in ESOL learner texts [C]// Proceedings of the Seventh Workshop on Building Educational Applications Using NLP, 2012: 33-43.

[7] Briscoe Y. F. H. Y. T. Neural Automated Essay Scoring and Coherence Modeling for Adversarially Crafted Input [C]// Proceedings of NAACL-HLT, 2018: 263-271.

[8] He Y., Jiang F., Chu X., et al. Automated Chinese Essay Scoring from Multiple Traits [C]// Proceedings of the 29th International Conference on Computational Linguistics, 2022: 3, 007-3, 016.

[9] Kenton J. D. M. W. C. and Toutanova L. K. BERT: Pre-training of Deep Bidirectional Transformers for Language Understanding [C]// Proceedings of NAACL-HLT, 2019: 4, 171-4, 186.

[10] Chen M., Goel K., Sohoni N. S., et al. Mandoline: Model evaluation under distribution shift [C]// International Conference on Machine Learning, PMLR, 2021: 1, 617-1, 629.

[11] Wang J., **Xu H. U., Hou W., et al. On the Robustness of ChatGPT: An Adversarial and Out-of-distribution Perspective [C]// ICLR 2023 Workshop on Trustworthy and Reliable Large-Scale Machine Learning Models, 2023.

# Research on $PM_{2.5}$ concentration prediction based on transformer

**Zhiqiang Dai [1,2], Ge Ren [2,3*], Yongming Jin [1,2] and Jintao Zhang [1,2,3]**

[1] School of Computer and Artificial Intelligence, Zhengzhou University, Zhengzhou, 450001, China
[2] Zhengzhou Institute of Metrology, Zhengzhou, 450001, China
[3] National Institute of Metrology, Beijing, 100029, China

* Corresponding author's E-mail: reng@nim.ac.cn

**Abstract:** The pollution caused by $PM_{2.5}$ can have serious consequences for human health and the environment, making timely prediction of its concentration important. This study proposes a Spatial-Temporal Air quality Transformer (STAFormer) for forecasting $PM_{2.5}$ concentration. In the dimension of time, this study designs a novel multi-head attention mechanism that can discover multi-periodicity in time series, thereby extracting more complex temporal features. In terms of spatial analysis, this study designs a graph convolutional neural network that incorporates distance and wind direction information to account for the impact of surrounding cities on $PM_{2.5}$ concentration prediction. Additionally, it dynamically captures spatial correlations using an attention mechanism. The prediction results for Zhengzhou show that the model suggested in this study performs better. The RMSE for forecasting $PM_{2.5}$ levels for the next hour is 3.51 $\mu g/m^3$, and MAE is 2.38 $\mu g/m^3$.

## 1. Introduction

Air quality is vital for our well-being, as high-quality air contributes to human health. $PM_{2.5}$ denotes fine particulate matter suspended in the air with a diameter of 2.5 μm or less, primarily from industrial emissions and vehicular exhaust. Long-term exposure to elevated levels of $PM_{2.5}$ can elevate the likelihood of cardiovascular diseases, respiratory ailments, and cancer, posing serious threats to human health. Predicting $PM_{2.5}$ concentration aids in early warning of environmental pollution, enabling governments to take timely measures and the public to plan outdoor activities accordingly, thus safeguarding public health. In summary, timely and precise prediction of $PM_{2.5}$ concentration is of paramount importance.

At present, $PM_{2.5}$ prediction mainly employs two types of methods: numerical models and statistical models. Numerical models utilize mathematical and physical processes to simulate pollution transport, dispersion, chemical reactions, etc., in the atmosphere. Based on principles of atmospheric science and meteorology, these models perform numerical simulations using computers to predict future air quality. Common numerical models include the CMAQ model, WRF-Chem model, and CAMx model, among others. The quality of input data influences the accuracy and reliability of these models, the selection of model parameters, and the accuracy of internal physical and chemical processes. Moreover, these models are computationally intensive and require long run times.

Statistical models mainly employ mathematical and statistical methods to establish models based on historical pollutant concentrations and meteorological data to forecast forthcoming air pollutant levels. Early on, traditional statistical methods such as multiple linear regression, ARIMA [1], etc., were predominantly used. Due to the simplicity of these methods, they fail to capture the hidden nonlinear relationships in the data. Subsequently, machine learning methods such as Support Vector Machine, Random Forest, XGBoost [2], etc., are beginning to be used for predicting $PM_{2.5}$ concentrations. As deep learning advances rapidly, many scholars have started employing deep learning models to forecast pollutant concentrations. Wen et al. [3] proposed a spatiotemporal convolutional LSTM extended model (C-LSTME) to predict air quality. Qi et al. [4] introduced a model integrating GCN and LSTM networks for predicting $PM_{2.5}$ concentrations. Pak et al. [5] presented a CNN-LSTM for predicting the next-day average $PM_{2.5}$ concentration. Xiao et al. [6] proposed DP-DDGCN to predict air quality. They first created a dual-path-directed graph based on the relationships between stations to capture spatial relationships and then used GRU to capture temporal correlations.

Most of the models mentioned above are based on LSTM models to capture the temporal dependencies in $PM_{2.5}$ concentration prediction. However, LSTM models may encounter the problem of gradient vanishing or explosion when dealing with long sequences, making it difficult to capture long-term dependencies. Transformer [7] utilizes attention mechanisms, enabling them to consider all positions in a sequence simultaneously and allowing for better capturing of long-term dependencies in time series. Additionally, the multi-head attention mechanism in Transformer models can attend to information at different scales, facilitating the capture of hourly, daily, and seasonal variations in $PM_{2.5}$ concentration. Furthermore, Transformers can be combined with other techniques, such as CNN and GCN, to capture spatial relationships and improve performance. These features of the Transformer model give it an advantage in predicting $PM_{2.5}$ concentrations. Therefore, this study utilizes it for $PM_{2.5}$ concentration prediction, adding temporal and spatial modules to extract spatiotemporal features better. Firstly, one-dimensional data is transformed into two-dimensional data by considering multiple cycles. Then, the traditional multi-head attention mechanism is modified to extract features from the two-dimensional data, thereby establishing the temporal module. Spatial relationships are captured through GCN, incorporating distance and wind direction information into the GCN to consider the impact of surrounding cities on $PM_{2.5}$ concentration prediction. Additionally, attention mechanisms are incorporated into the graph neural network to account for the influence of neighboring cities on dynamic prediction. Finally, the STAFormer model is developed.

## 2. Materials and methods

### *2.1. Data source*

The data we utilized consisted of air quality (https://quotsoft.net/air/) and meteorological data (https://cds.climate.copernicus.eu). As shown in Figure 2, we collected air quality data from 312 cities. The air quality data for each city is obtained by averaging the data from all national monitoring stations. For each city mentioned above, we collected corresponding meteorological data. Table 1 lists the variables used in this investigation. We standardized the data using the Z-score normalization method.

**Table 1.** Input variables utilized in this study.

| Variable Name | Unit |
|---|---|
| $PM_{2.5}$ | μg/m³ |
| CO | mg/m³ |
| $NO_2$ | μg/m³ |
| $O_3$ | μg/m³ |
| $SO_2$ | μg/m³ |
| $PM_{10}$ | μg/m³ |
| Surface_pressure | Pa |
| Temperature_2 m | K |
| Total_precipitation | m |
| Dewpoint_temperature | K |
| U_component_of_wind_10 m | m/s |
| V_component_of_wind_10 m | m/s |

*2.2. The Model Structure*

We made improvements based on the Transformer model. The Transformer model follows an encoder-decoder architecture, where for an input sequence $(x_1, \ldots, x_n)$, the encoder maps it to $z = (z_1, \ldots, z_n)$, and then the decoder utilizes z to generate the output sequence $(y_1, \ldots, y_n)$ step by step, in an autoregressive manner. Figure 1 depicts the structure of STAFormer. Compared to the Transformer, it has been modified primarily in two aspects. Firstly, a time preprocessing module is added before the multi-head attention mechanism, converting the one-dimensional data into two-dimensional data considering multiple periods, with the addition of two-dimensional convolution in the multi-head attention mechanism to extract two-dimensional features. Secondly, the position-wise feedforward network is replaced with GCN to extract spatial features of pollutant concentration. Both the encoder and decoder comprise N identical layers. We only made modifications to the multi-head attention in the encoder. The output dimension of all sub-layers is set to $d_{model}$. In this experiment, we set N=3 and $d_{model}$=64.

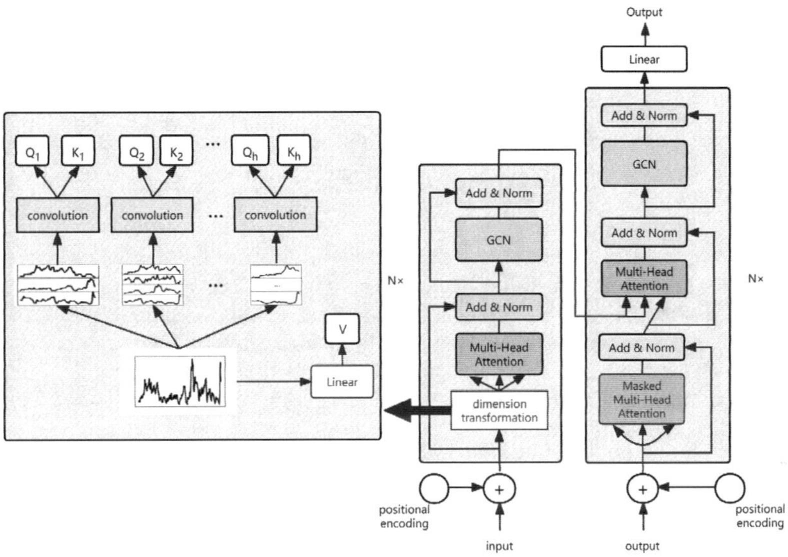


**Figure 1.** STAFormer architecture.

*2.3. Temporal module*

Time series data frequently display multiple periodic patterns, such as daily and monthly variations. The variation at a specific time point within one period is influenced by neighboring times within the same period and adjacent periods. For instance, concerning $PM_{2.5}$ concentration, at any given moment, the concentration value may be influenced by values at neighboring moments and the concentration value at the same moment on the preceding day. Therefore, time series can be divided into multiple periods to capture changes within and between periods. This transforms the one-dimensional time series into two-dimensional data, with rows representing within-period changes and columns representing between-period changes. Following the approach of the TimesNet model [8], we implement this operation using Equation (1):

$$\{f_1, \ldots, f_k\} = TopkFre\left(Amplitude\left(FFT(X)\right)\right),$$
$$p_i = \left\lceil \frac{1}{f_i} \right\rceil, i \in \{1, \ldots, k\} \quad (1)$$

where FFT(·) represents the Fast Fourier Transform, Amplitude(·) calculates the amplitude, and TopkFre(·) selects the frequencies $f_i$ corresponding to the top k largest amplitudes. Firstly, the frequency domain is decomposed through the Fast Fourier Transform, and then k frequencies corresponding to the highest amplitudes are selected, thus obtaining the period $p_i$ corresponding to each frequency. By using $p_i$, the original sequence T is transformed into $(T/p_i, p_i)$, effectively converting the one-dimensional sequence into two-dimensional data.

Modifications are made to the original Transformer model's multi-head attention mechanism. Query and Key are obtained through linear mappings in the traditional attention mechanism. However, in this study, Query and Key are obtained through two-dimensional convolution on the input sequence, enabling better feature extraction from the previously generated two-dimensional data, as shown in Equation (2).

$$MultiHead(Q,K,V) = Concat(head_1, \dots, head_h)W^O$$
$$where\ head_i = Attention(Q_i, K_i, V_i)$$
$$where\ Q_i = X * \Phi_i^Q, K_i = X * \Phi_i^K, V_i = XW_i^V \tag{2}$$

where X is the input sequence after undergoing periodic transformation, Q, K, and V represent Query, Key, and Value, respectively. $head_i$ denotes the output of the i-th attention head, $W^O$ is the weight matrix used to concatenate and linearly transform the outputs of multiple heads. $Q_i$, $K_i$ and $V_i$ represent the Q, K, and V corresponding to the i-th attention head. $\Phi_i^Q$ and $\Phi_i^K$ are the convolutional kernels, with the kernel size set to 3×3. It can be seen that $Q_i$ and $K_i$ are obtained through convolutional operations on X.

### *2.4. Spatial module*

The concentration of $PM_{2.5}$ is not only influenced by historical $PM_{2.5}$ concentration data but also by neighboring cities. This study incorporates GCN to account for the influence of neighboring cities on Zhengzhou. We collected data from a total of 312 cities. We calculated the Pearson correlation coefficient and mutual information between $PM_{2.5}$ concentrations of each city and Zhengzhou City. The top 50 cities in both measures were selected, resulting in the selection of 48 cities in addition to Zhengzhou City. Thus, data from 49 cities were input into the model, as shown in Figure 2.

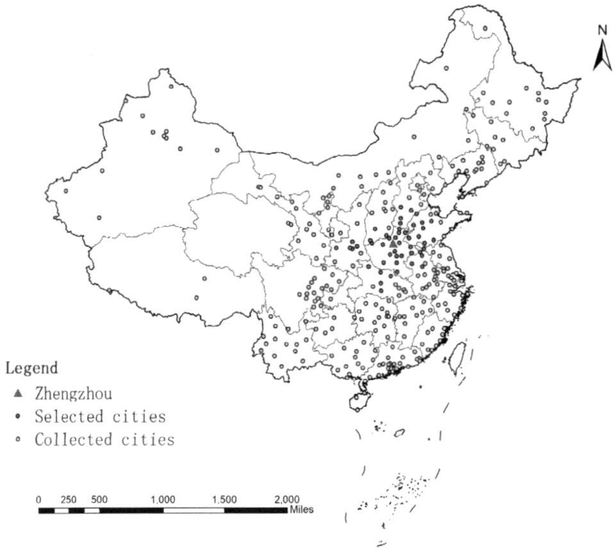


**Figure 2.** Selected cities.

We establish a graph $G = (V, E)$. V represents the 49 selected cities mentioned above. Regarding edges E, undirected edges are created between Zhengzhou City and the other 48 cities. For the other cities, undirected edges are created solely with the top five cities based on the Pearson correlation coefficient, given that the coefficient exceeds 0.6. The constructed graph can be represented by an adjacency matrix A. For every element in the matrix, 1 indicates the presence of an edge between the two cities, while 0 indicates no edge between them.

This study captures spatial correlations through GCN and modifies it to consider the influence of distance and wind direction on pollutant concentration. Inspired by ASTGNN [9], this study also employs attention mechanisms to compute the dynamic correlations between different cities automatically. The implementation of GCN is shown in Equation (3),

$$GCN\left(Z^{(l)}\right) = \sigma\left(\tilde{A}Z^{(l-1)}W^{(l)}\right) \tag{3}$$

where $Z^{(l)}$ represents the output of the l-th layer of GCN. $W^{(l)}$ denotes the weight matrix of the l-th layer of GCN, used for linear transformation of input features. σ represents the activation function, and $\tilde{A}$ denotes the normalized adjacency matrix.

To consider the impacts of distance and wind direction on $PM_{2.5}$ concentration prediction, the distance and wind direction information are integrated into the matrix A. For distance, we introduce a distance weight matrix $S_1$, defined as the reciprocal of the distance between different nodes. Then, $S_1$ is multiplied element-wise with $\tilde{A}$ to adjust the matrix $\tilde{A}$, thereby introducing distance-related correlations into our spatial module, as shown in Equation (4). The influence of stations farther away from Zhengzhou city will diminish, while the weights of stations closer to Zhengzhou will increase.

For wind direction, we introduce a wind direction weight matrix $S_2$, as shown in Equation (5) and (6). First, the directional matrix $D_2$ between each city and Zhengzhou city is computed. Then, $D_2$ is subtracted from the wind direction α of Zhengzhou city, and the resulting matrix is divided by 45 and then reciprocated to obtain the final wind speed weight matrix $S_2$. This effectively divides the cities surrounding Zhengzhou city into 4 sectors based on direction. If at a certain moment, a city's direction relative to Zhengzhou City aligns closely with the wind direction in Zhengzhou City, then the influence of that city on Zhengzhou City is greater. By performing element-wise multiplication between $S_2$ and $\tilde{A}$, we account for the influence of different cities on Zhengzhou city regarding wind speed. Equations (4), (5) and (6) are as follows.

$$S_1 = \frac{1}{D_1} \tag{4}$$

$$D_3 = min(|D_2 - \alpha|, 360 - |D_2 - \alpha|) \tag{5}$$

$$S_2 = \frac{1}{\left\lfloor \frac{D_3 - 1}{45} \right\rfloor + 1} \tag{6}$$

where $D_1$ represents the distance matrix between nodes, $D_2$ represents the direction of each node relative to the Zhengzhou node, and α represents the wind direction in Zhengzhou at a certain moment.

Considering that the correlation between nodes may change over time, we utilize an attention mechanism to dynamically compute the correlation between nodes, introducing the spatial correlation weight matrix $S_3$, as shown in Equation (7). This approach allows us to consider the dynamics of spatial relationships and enables the model to learn the underlying relationships between nodes automatically.

$$S_3 = softmax\left(\frac{Z^{(l-1)}Z^{(l-1)^T}}{\sqrt{d_{model}}}\right) \epsilon R^{N\times N} \tag{7}$$

where $d_{model}$ is the dimensionality of queries and keys, $Z^{(l-1)}$ represents the output before the GCN sublayer, and $S_{ij}$ in $S_3$ denotes the correlation strength between node $i$ and node $j$, with larger values indicating stronger correlation. We adjust $\tilde{A}$ using the spatial correlation weight matrix $S_3$ through element-wise dot product operation. Finally, the three modules mentioned above are combined, as shown in Equation (8),

$$X^{(l)} = \sigma\left(\frac{1}{3}\sum_{k=1}^{3}(\tilde{A}\odot S_k)Z^{(l-1)}W^k\right) \tag{8}$$

*2.5. Settings*

The dataset includes air quality and meteorological data from 49 selected cities from January 1, 2021, to August 31, 2023, with hourly temporal resolution. Data from January 2021 to December 2022 are employed as the training set, data from January to April 2023 are used as the validation set, and data from May to August 2023 are used as the test set. The model is optimized using Adam. We evaluate the model using two metrics, MAE and RMSE, as defined in Equations (9) and (10), respectively.

$$MAE = \frac{1}{n}\sum_{i=1}^{n}|y_i - \hat{y}_i| \tag{9}$$

$$RMSE = \sqrt{\frac{1}{n}\sum_{i=1}^{n}(y_i - \hat{y}_i)^2} \tag{10}$$

Here $y_i$ represents the true value, and $\hat{y}_i$ represents the model-predicted value. The smaller these two metrics are, the better the model's performance is. Each experiment is conducted three times, and the average of the results from these three experiments is considered the final evaluation.

The study was conducted on a single T4 GPU, implementing the model using PyTorch. MAE was chosen as the loss function. The model dimension ($d_{model}$) was configured to 64, the number of attention heads (h) was set to 8, the batch size was set to 32, and the learning rate was set to 0.001. To confirm the effectiveness of the proposed model, we compared it with MLP, LSTM, and $PM_{2.5}$-GNN. $PM_{2.5}$-GNN [10] is a model tailored for the prediction of $PM_{2.5}$. It utilizes GNN to incorporate domain knowledge and integrates with RNN to capture long-term dependencies in the $PM_{2.5}$ process.

## 3. Analysis of experimental results

This study predicts the $PM_{2.5}$ concentration for the next 6 hours using historical data from the past 24 hours. Table 2 displays the results, with the optimal results emphasized in bold. It can be observed that both RMSE and MAE of our model outperform the baseline models, particularly in the first three hours. As depicted in Figure 3, overall, the RMSE and MAE of the four models increase with the length of the forecast horizon. The RMSE values for MLP, LSTM, $PM_{2.5}$-GNN, and STAFormer predicting 1 hour are 8.8094 μg/m$^3$, 7.7243 μg/m$^3$, 8.3505 μg/m$^3$, and 3.5184 μg/m$^3$, respectively, while the MAE values are 6.3116 μg/m$^3$, 5.5811 μg/m$^3$, 7.3687 μg/m$^3$, and 2.3806 μg/m$^3$, respectively. When predicting the next hour, STAFormer's RMSE decreased by 60.06%, 54.45%, and 57.87% compared to MLP, LSTM, and $PM_{2.5}$-GNN, respectively, while MAE decreased by 62.28%, 57.35%, and 67.69%, respectively. When predicting the preceding 6 hours, RMSE decreased by 22.34%, 17.21%, and 22.37% compared to MLP, LSTM, and $PM_{2.5}$-GNN, respectively, while MAE decreased by 27.15%, 21.39%, and 31.57%, respectively.

**Table 2.** The results of different model.

| Model | Metric | 1 h | 2 h | 3 h | 4 h | 5 h | 6 h |
|---|---|---|---|---|---|---|---|
| MLP | RMSE | 8.8094 | 9.5542 | 10.0054 | 10.3725 | 10.7099 | 11.1369 |
| | MAE | 6.3116 | 6.9216 | 7.3019 | 7.6202 | 7.8906 | 8.2223 |
| LSTM | RMSE | 7.7243 | 8.4502 | 9.2275 | 9.9211 | 10.5094 | 11.0056 |
| | MAE | 5.5811 | 6.0757 | 6.6220 | 7.1432 | 7.6003 | 8.0019 |
| $PM_{2.5}$-GNN | RMSE | 8.3505 | 8.8413 | 9.4901 | 10.4020 | 11.3570 | 12.1724 |
| | MAE | 7.3687 | 7.5635 | 7.6123 | 7.9205 | 8.2190 | 8.4410 |
| STAFormer | RMSE | **3.5184** | **5.7552** | **7.6115** | **9.0358** | **10.1330** | **11.0004** |
| | MAE | **2.3806** | **3.8359** | **5.1036** | **6.1320** | **7.0251** | **7.7701** |

MLP is a relatively simple model, which, in this experiment, predicts future $PM_{2.5}$ concentrations based on historical data. MLP does not consider spatial relationships in $PM_{2.5}$ concentration prediction nor design specific structures to consider temporal relationships. From the experimental results, it can be observed that MLP performs worse than LSTM and STAFormer. The LSTM model considers temporal relationships but does not consider spatial relationships, which are also crucial for predicting $PM_{2.5}$ concentrations. The performance of the LSTM is superior to MLP but lags behind STAFormer, which considers both temporal and spatial relationships.

$PM_{2.5}$-GNN model is specifically tailored for predicting $PM_{2.5}$ concentrations. However, this model primarily relies on forecasted meteorological data while overlooking the utilization of historical air quality data, which is also crucial. Therefore, the performance of the $PM_{2.5}$-GNN model is constrained by the accuracy of the forecasted meteorological data. Since obtaining accurate meteorological forecast data is challenging, the $PM_{2.5}$-GNN model performs poorly in this experiment. However, $PM_{2.5}$-GNN indeed has a significant advantage in extracting spatial relationships. If MLP and LSTM models also rely on forecasted meteorological data and the $PM_{2.5}$ data from the previous time step obtained by the model for training, similar to $PM_{2.5}$-GNN, then the performance of the $PM_{2.5}$-GNN model would surpass that of the MLP and LSTM models.

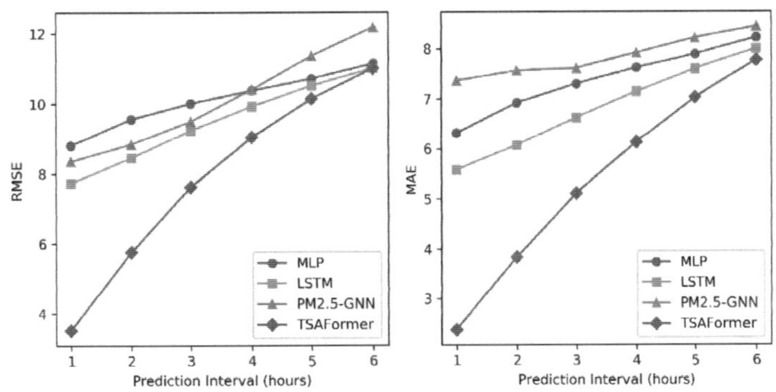


**Figure 3.** Performance changes of different methods as the prediction interval increases.

## 4. Conclusions

This study proposed STAFormer to predict $PM_{2.5}$ concentrations. The model is designed to model both temporal and spatial aspects. In terms of time, we designed a special multi-head attention mechanism that transforms the one-dimensional sequence into two-dimensional data, considering intra-period and inter-period segments. Then, Q and K are generated through two-dimensional convolution, followed by multi-head attention operations. Regarding spatial analysis, we modified the GCN to capture spatial correlations while considering the influence of distance and wind direction. Additionally, we introduced a dynamic module to adjust the spatial correlation strength adaptively. Experimental results for predicting $PM_{2.5}$ concentrations in Zhengzhou demonstrate that this model outperforms the baseline. In summary, this study considers the spatiotemporal characteristics. Although this study focuses only on predicting the $PM_{2.5}$ concentration in Zhengzhou, the analytical process can also be extended to predict pollutant concentrations in other cities.

## Acknowledgment

This study was supported by the Fundamental Research Funds for Central Public Welfare Scientific Research Institutes sponsored by the National Institute of Metrology, China (AKYZD2207-2) and the National Key Technologies Research and Development Program of China (2019YFF0217002).

**References**

[1] Zhang L, Lin J, Qiu R, et al. Trend analysis and forecast of PM2. 5 in Fuzhou, China using the ARIMA model [J]. Ecological Indicators, 2018, 95: 702-710.

[2] Ma J, Yu Z, Qu Y, et al. Application of the XGBoost machine learning method in PM2.5 prediction: A case study of Shanghai [J]. Aerosol and Air Quality Research, 2020, 20(1): 128-138.

[3] Wen C, Liu S, Yao X, et al. A novel spatiotemporal convolutional long short-term neural network for air pollution prediction [J]. Science of the total environment, 2019, 654: 1091-1099.

[4] Qi Y, Li Q, Karimian H, et al. A hybrid model for spatiotemporal forecasting of PM2.5 based on graph convolutional neural network and long short-term memory [J]. Science of the Total Environment, 2019, 664: 1-10.

[5] Pak U, Ma J, Ryu U, et al. Deep learning-based PM2.5 prediction considering the spatiotemporal correlations: A case study of Beijing, China [J]. Science of the Total Environment, 2020, 699: 133561.

[6] Xiao X, Jin Z, Wang S, et al. A dual-path dynamic directed graph convolutional network for air quality prediction [J]. Science of The Total Environment, 2022, 827: 154298.

[7] Vaswani A, Shazeer N, Parmar N, et al. Attention is all you need [J]. Advances in neural information processing systems, 2017, 30.

[8] Wu H, Hu T, Liu Y, et al. Timesnet: Temporal 2d-variation modeling for general time series analysis [J]. arXiv preprint arXiv:2210.02186, 2022.

[9] Guo S, Lin Y, Wan H, et al. Learning dynamics and heterogeneity of spatial-temporal graph data for traffic forecasting [J]. IEEE Transactions on Knowledge and Data Engineering, 2021, 34(11): 5415-5428.

[10] Wang S, Li Y, Zhang J, et al. PM2. 5-GNN: A domain knowledge enhanced graph neural network for PM2.5 forecasting [C]//Proceedings of the 28th international conference on advances in geographic information systems. 2020: 163-166.

Journal of Physics: Conference Series **2813** (2024) 012024 doi:10.1088/1742-6596/2813/1/012024

# A time series data regression based on a B-spline and stacking framework

**Ruizhe Li[1], Jingyu Cui[2] and Daobo Fu[3*]**

[1]Sun Yat-sen University, Guangzhou, China
[2]Shandong University of Science and Technology, Shandong, China
[3]Anhui University of Finance and Economics, Anhui, China
Ruizhe Li, Jingyu Cui and Daobo Fu contributed equally to this work

[*]Corresponding author's e-mail: 762316118@qq.com

**Abstract:** This paper proposes a new time series regression method based on the combination of B-spline basis expansion and distance covariance weighted model. This paper considers the case that the predictor variable is a time series and the response variable is a continuous scalar. In order to solve these problems of high dimension and high noise of time series, this proposed method uses a B-spline basis to reduce the dimension and extract the function information of the original sequence. Considering that there is still some noise in the reduced dimension random variables, this proposed method considers using the stacking framework to enhance the data. Since the importance of each basis model is unknown, this paper uses the distance covariance to measure the importance of the basis model and develop a weighted stacking framework, which can adaptively weigh the bias and variance of the predictor model. In addition, each basis estimator in ensemble learning is model-free. The real data analysis shows that the proposed method is competitive with some existing methods. Finally, the proposed method can be directly applied to other fields such as environmental science, medical science and speech recognition et al.

## 1. Introduction

Time series analysis is an important branch of statistical science. It mainly studies the process of occurrence and development of things with the change of time, looks for the laws of development and change of things, and forecasts the future trend. At present, time series analysis is widely used in many fields, such as economy, finance, astronomy, meteorology, ocean, and so on. Time series data often has the following characteristics. On the one hand, time series data tend to have autocorrelation; on the other hand, due to too many factors affecting response variables in actual production and life, the number of predictor variables is too large, and it is easy to have the problem of "dimension disaster". This paper considers the case where the predictor variable is a time series and the response variable is a continuous random variable, that is, a time series data regression problem. Next, this paper uses the meat sample data set to further explain the data background. This data set consists of 100 channels of absorbance spectra and the contents of moisture (water), fat, and protein. The absorbance is -log10 of the transmittance measured by the spectrometer. In addition, the contents of water, fat, and protein are measured by specific chemical analysis methods and calculated in percentage. Our goal is to predict the content of water, fat, and protein based on the time series data formed by absorbance.

In statistics, for regression problems, on the one hand, classical linear models [1] include but are not limited to simple linear regression, multiple linear regression, polynomial linear regression, ridge regression, lasso regression, etc. On the other hand, there are many nonlinear models to deal with nonlinear regression problems [2], such as regression trees, logistic regression, etc. In addition, there are integrated learning models with a wide range of applications. For example, Huang et al. [3] studied the track simulation of tropical cyclones in the western North Pacific using a random forest model, Nguyen et al. [4] predicted the swelling potential of soil using a combination of bagging algorithm and multi-layer perceptron hybrid intelligence model, and Xie et al. [5] used boosting algorithm to denoise the image. However, under the background of time series data regression, the above methods did not solve the problems of dimensional disaster on the one hand. On the other hand, they may retain or even amplify the irrelevant noise, so that the prediction accuracy is greatly reduced. The most commonly used method to solve this problem is two-stage regression. Specifically, the dimension reduction method is first used to process the time series data, and then the classical regression model is used to predict the response variables. The most famous linear dimension reduction method is principal component analysis (PCA), and other methods include CVA, KPCA, etc. For example, Feng et al. [6] combined the PCA dimension reduction method and logistic regression method to predict pneumonia data in medicine, while Xu et al. [7] used PCA and deep forest regression [8] to predict dioxin emission concentration. Montalvo et al. [9] used PCA and linear regression to predict the monthly electricity consumption of healthcare buildings, and Perera et al. [10] used PCA and partial least square regression (PLS-R) combined with ATR-FTIR spectroscopy to study the concentration-dependent curcumin interaction with serum biomolecules. However, this kind of method regards each observation time point as a one-dimensional variable and does not take into account the intrinsic function characteristic information of time series data. Moreover, because of the strong correlation between time points, the regression model based on the multivariate dimension reduction method has limitations. Based on this, in order to extract the functional features of time series and avoid the influence of correlation between observation points, a method called basis expansion can be used. According to the form and nature of the basis function, it can be divided into multiple sub-methods, such as B-spline basis expansion, smooth spline expansion, Fourier basis expansion, etc. Compared with the traditional linear dimension reduction method, the basis expansion methods have the advantage of better extracting the function information of the original sequence. For example, Boukezzoula et al [11] used B-spline expansion and possibilistic regression to predict lake sediments, Bignardi et al. [12] combined B-spline expansion and random regression model (RR) to predict test-day milk yield of Holstein cattle in Brazil, and Kazuo et al [13] used third-degree B-spline expansion and self-modeling regression to analyze functional data of mandibular movement. Soh and Zhu [14] used the Fourier basis expansion and lasso regression model to predict the purity of olive oil blends. However, such methods did not take into account the potential model structure between the base expansion coefficient variables and the response variables. In addition, although the methods of basis function expansion achieve the purpose of dimension reduction and extract the function features, the noise still exists, and because of the loss of information after dimension reduction, how to enhance the data is also a problem.

In order to solve the above problems, this paper proposes a time series regression method combining B-spline basis expansion and distance covariance weighted model. The main contributions of the proposed method are as follows: firstly, the smoothing method is used to deal with the problems of high dimension and large noise of time series data, and the function information of time series data is extracted. Secondly, the stacking framework is used for data enhancement, which increases the effective information weight of response variables and reduces the negative impact of interference noise. Thirdly, a model averaging method based on distance covariance weighting is developed to weigh the bias and variance of the prediction model, and then improve the prediction accuracy. In addition, the regression method in the base estimator is model-free. The analysis of actual data shows that the proposed method is competitive with some existing methods. Finally, the proposed method can be directly applied to other fields such as environmental science, medical science, and speech recognition. The rest of this paper is arranged as follows: Section 2 introduces the time series regression model based on the fusion of the smoothing method and distance covariance weighted model; Section 3 shows the actual data analysis results; Section 4 is the summary.

## 2. Theory and concept

### *2.1 Sample base expansion*

It is assumed that $n+1$ control points $P_i(i=0,1,2,\cdots,n)$, and a node vector $U=\{u_0,u_1,\cdots,u_m\}$, pth B-sample curve is defined by these control points and node vector $U$. We make $N_{i,p(u)}$ be the ith pth B−sample basis function, and the pth B−sample curve formula can be defined as:

$$C(u)=\sum_{i=0}^{n} N_{i,p}(u)P_i \quad (1)$$

There are various types of B-sample basis functions that combine into B-sample curves, of which the application of deBoor-cox defined basis functions is widely used. We make $U=\{u_0,u_1,\cdots,u_m\}$ be a monotonically nondecreasing sequence of real numbers, $u_i \le u_{i+1}, i=0,1,\cdots,m-1$. $U$ is a node vector, and the half-open interval $[u_i,u_{i+1})$ is the i-th node interval, then the basis functions are shown as follows:

$$N_{i,p}(u)=\begin{cases}1 & \text{if } u_i \le u < u_{i+1}\\ 0 & \text{otherwise}\end{cases}$$

$$N_{i,p}(u)=\frac{u-u_i}{u_{i+p}-u_i}N_{i,p-1}(u)+\frac{u_{i+p+1}-u}{u_{i+p+1}-u_{i+1}}N_{i+1,p-1}(u) \quad (2)$$

If $p=0$, the basis function $N_{i,p(u)}$ is a step function in the interval $[u_i,u_{i+1})$. When $p>0$, $N_{i,0}$ is a linear combination of two $p-1$th subbase functions. The B-sample curve is obtained by fitting each data with the basis expansion method on the node vector $U=\{u_0,u_1,\cdots,u_m\}$, which further gives the B-sample coefficient vector.

### *2.2 Stacking model fusion and weighted bagging framework*

Since there is still noise in the time series after dimension reduction, the stacking framework can be used to enhance the important features in the original data to achieve data enhancement. Stacking model fusion combines several different base models by learning them and training a metamodel to combine them, and then outputs a final prediction based on the multiple predictions returned by these base models. When the base models are combined correctly, the model with better robustness and generalization capabilities can be obtained. In addition, the bagging (bootstrap aggregating) framework selects homogeneous base learners, allows the base learners to train independently of each other in parallel, and then combines the results of all base learners. A total of k training samples are randomly drawn from the original sample set, and a total of n rounds of drawing are performed to obtain n training sets (the n training sets are independent of each other, and elements can have duplicates). For the n training sets, k models are trained. For the classification problem, the classification result is generated by voting; for the regression problem, the mean of the k models' prediction results is used as the final prediction. Since the importance of the base models is unknown, this study assigns corresponding weights according to the importance of different base models to obtain a combination of models with better performance. Distance Covariance ($dcov(u,v)$) and Distance Correlation ($dcorr(u,v)$) are measures of correlation between two variables, and the distance correlation coefficient is chosen in this study to assign weights to the base model in stacking. The distance correlation coefficients are calculated as follows:

$$\hat{d}corr(u,v)=\frac{\hat{d}cov(u,v)}{\sqrt{\hat{d}cov(u,u)\hat{d}cov(v,v)}} \quad (3)$$

where $\hat{d}cov^2(u,v)=\hat{s}_1+\hat{s}_2-2\hat{s}_3$, $\hat{s}_1,\hat{s}_2,\hat{s}_3$ are: $\hat{s}_1=\frac{1}{n^2}\sum_{i=1}^{n}\sum_{j=1}^{n}\|u_i-u_j\|_{d_u}\|u_i-u_j\|_{d_v}\hat{s}_2=\frac{1}{n^2}\sum_{i=1}^{n}\sum_{j=1}^{n}\|u_i-u_j\|_{d_u}\frac{1}{n^2}\sum_{i=1}^{n}\sum_{j=1}^{n}\|u_i-u_j\|_{d_v}$; $\hat{s}_3=\frac{1}{n^2}\sum_{i=1}^{n}\sum_{j=1}^{n}\sum_{l=1}^{n}\|u_i-u_j\|_{d_u}\|u_i-u_j\|_{d_v}$. Similarly, $dcov(u,u)$ and $dcov(v,v)$ are calculated. The larger the correlation $I_i(I=1,2,\cdots,j)$ is, the greater the contribution of the model is to obtaining better predictions, assigning a higher weight to it. The weight of the base model is $w_j$.

### *2.3 Time series regression model*

This paper proposes a time series regression model (BSDMW) based on the fusion of B spline basis expansion and distance correlation coefficient weighting model to solve the problems of dimensional catastrophe and underfitting that can be encountered in regression problems of time series. In addition, since time series can be viewed as continuous functions, important features in the functions need to be captured and dimensionality reduction is applied to high-dimensional data. To obtain better prediction results, this paper assigns weights to the base models according to the degree of importance of different base models. The specific algorithm is shown in Figure 1.

**Algorithm 1**

**input:** $X_T, Y_T, X_V, Y_V, X_e, \boldsymbol{\theta}_1, \boldsymbol{\theta}_2, L, \mathrm{K}$
**out:** $\hat{Y}_c$
$\{Z_T\}_{i=1}^{n_T}, \{Z_V\}_{i=1}^{n_V}, \{Z_e\}_{i=1}^{n_e} := Bspline((X_T, X_V, X_e), basisnumber = n_T + n_V + n_e)$
**for all** $k, \theta_1, \theta_2, l$ in $K, \boldsymbol{\theta}_1, \boldsymbol{\theta}_2, L$ **do**
  **for all** $s$ in $S$ **do**
    $\widehat{M}_s := M_s(Z_T^{(-s)}, Y_T^{(-s)}, \theta_1, k - fold = k)$
    $\left\{B_T^{(p)}, B_V^{(p)}\right\}_{p=1} := \widehat{M}_s(\{Z_T\}_{i=1}^{n_T}, \{Z_V\}_{i=1}^{n_V}, \theta_1, k - fold = k)$
    $\widehat{MA} := MA\left(\left\{B_T^{(-s)}, Y_T^{(-s)}\right\}, l, \theta_2\right)$
    $\left\{\hat{Y}_V^{(q)}\right\}_{q=1}^{l} := \widehat{MA}\left(\left\{B_V^{(p)}\right\}_{p=1}, l, \theta_2\right)$
    $\left\{dcov_V^{(q)}\right\}_{q=1}^{l} := DCOV\left(\left\{\hat{Y}_V^{(q)}\right\}_{q=1}^{l}, Y_V\right)$
    $\{W_q\}_{q=1}^{l} := \frac{\left\{dcov_V^{(q)}\right\}_{q=1}^{l}}{\sum_q (dcov_V^{(q)})}$
    $\left\{\hat{Y}_{T,q}^{(s)}\right\}_{q=1}^{l} := \widehat{MA}\left(\left\{B_T^{(p,s)}\right\}_{p=1}^{k}, l, \theta_2\right)$
    $\hat{Y_T}^{(s)} = W^{\top} \cdot \left\{\hat{Y}_{T,q}^{(s)}\right\}_{q=1}^{l}$
    $E_s = \sum_i \left(\hat{Y_T}^{(s)} - Y_T^{(s)}\right)^2$
  **end for**
**end for**
$k^*, \theta_1^*, \theta_2^*, l^*, w^* := \mathrm{argmin}_{k,\theta_1,\theta_2,L,w} \frac{1}{S} \sum_{s=1}^{S} E_s$
**return** $\hat{Y}_c := (W^*)^{\top} \cdot \widehat{MA}\left(\widehat{M}_s(\{z_c\}_{:=1}^{n}, \theta_1^*, k^*), l^*, \theta_2^*\right)$

**Figure 1.** Pseudo-code diagram of BSDMW mode.

We define the training set as $\{X_T, Y_T\} = (x_i^T, y_i^T)_{\mathrm{i=1}}^{n_T}$, the validation set as $\{X_V, Y_V\} = (x_i^V, y_i^V)_{\mathrm{i=1}}^{n_V}$, and the test set $\{X_e, Y_e\} = (x_i^e, y_i^e)_{\mathrm{i=1}}^{n_e}$. The stacking framework is denoted as $M_S = (m_1^s, m_2^s, m_3^s, \cdots, m_J^s)$, where $m_j^s (1 \le j \le J)$ is the base model, and the set of hyperparameters of the base model is defined as $\theta_1$. The model averaging framework is defined as $\mathrm{MA} = (MA_1,\ MA_2, \mathrm{M3}, \cdots, \mathrm{MA}_L)$, where $\mathrm{MA}_l\ (1 \le l \le L)$ is the base model and the set of hyperparameters of the base model is defined as $\theta_2$. The set of cross-validation folds in the stacking framework is K, $\mathrm{k} \epsilon K$. We define the B spline basis coefficient estimate as Bspline, which divides $X_T, Y_T$ into S mutually disjoint uniform subsets defined as $\left\{(X_T^{(s)}, Y_T^{(s)})\right\}_{s=1}^{S}$, where $(X_T^{(-s)}, Y_T^{(-s)})$ denotes all samples except the $s$th subset. L is defined as the number of base models in the model averaging framework. It is assumed that the base models of the stacking framework and their numbers are known, DCOV is the distance covariance estimate, dcov is the distance covariance score, and W is the model weight vector.

## 3. Data analysis

### *3.1 Data sources and data characteristics*

The original dataset used in this section comes from a quality control problem in food processing, the data is named tecator and is directly available at **http://lib.stat.cmu.edu/datasets/tecator**. The tecator dataset contains 215 light absorption spectra observed on minced meat slices as a function of wavelength, i.e. sample size. These data were recorded on a Tecator Infratec food and feed analyser in the wavelength range 850-1050 nm by means of the Near Infrared Transmission (NIT) principle. For each sample, the data consists of 100 channels of absorbance spectra $(x_1, x_2, x_3, \cdots, x_{100})$ and numerical values for the content of fat ($y_1$), water ($y_2$), and protein ($y_3$). The absorbance is -log10 of the transmittance measured by the spectrometer. The three contents of fat, water and protein are expressed as percentages by chemical analysis.

### *3.2 Comparison models and programming environment and calculation*

This paper considered seven methods, namely Linear Regression Model (LR), KNN Regression Model (KNN), LASSO Regression Model (Lasso), Decision Tree Regression Model (Tree), AdaBoost Regression Model (Ada), Gradient Boosting Regression Model (GBDT) and Random Forest Regression Model (RF), and compared the prediction results of these methods with the BSDMW model proposed in this paper in the case of data dimensionality reduction and non-dimensionality reduction. The data dimensionality reduction methods we use are PCA and KPCA. Among them, PCA is a method to reduce data dimensionality with little loss of data set information through feature space transformation. KPCA method uses the idea of the kernel function, whose basic principle is to perform the nonlinear transformation on the sample and perform principal component analysis in the transformed space to realize the nonlinear principal component analysis in the original space. In order to measure the prediction accuracy of the above methods and evaluate the pros and cons of different models, this paper conducted 100 simulation experiments through these models respectively, calculated the absolute error (AE), the mean square error (MSE), and the adjusted $R^2$ ($AR^2$) of each estimator on the test set $S_{test} = \left\{\left(y_i^*, x_i^{*'}, e_i^*\right)', i = 1,2,\cdots,n\right\}$ and calculated the mean and standard deviation of them as evaluation metrics. The smaller AE and MSE are, the larger $AR^2$ is, which indicates the higher prediction accuracy. The definitions of the three are as follows:

$$\mathrm{MSE}_j = \frac{1}{n}\sum_{i=1}^{n}\left(\hat{y}_{j,i} - y_i\right)^2, \mathrm{AE}_j = \sum_{i=1}^{n}\left|\hat{y}_{j,i} - y_i\right|, AR^2{}_j = 1 - (1 - R^2)\frac{n-1}{n-k-1} \quad (4)$$

where $j$ represents the above seven comparison models and the BSDMW model under the condition of data dimensionality reduction and non-dimensionality reduction, $j = 1,2,\cdots,22$. $y_i$ and $\hat{y}_{j,i}$ represent the actual and predicted values of the corresponding explanatory variables when model $j$ is used at time $i$, respectively.

**Table 1.** Fitting accuracy of 100 experimental results (fat content of meat samples).

| Method | AE | MSE | AR² | Method | MSE | AE | AR² |
|---|---|---|---|---|---|---|---|
| **Ada** | 7.45 (0.63) | 80.83 (14.05) | 0.49 (0.08) | **KNN** | 7.09 (0.82) | 84.61 (20.11) | 0.47 (0.10) |
| **PCA-Ada** | 7.45 (0.64) | 80.16 (13.09) | 0.50 (0.08) | **PCA-KNN** | 8.73 (0.87) | 130.96 (24.85) | 0.19 (0.11) |
| **KPCA-Ada** | 9.70 (0.70) | 140.9 (19.10) | 0.14 (0.08) | **KPCA-KNN** | 9.85 (0.70) | 153.77 (22.05) | 0.12 (0.09) |
| **Tree** | 11.99 (1.23) | 248.0 (44.00) | 0.57 (0.32) | **Lasso** | 9.05 (0.68) | 122.05 (19.00) | 0.24 (0.08) |
| **PCA-Tree** | 10.41 (1.31) | 216.8 (44.07) | 0.38 (0.26) | **PCA-Lasso** | 9.47 (0.65) | 132.87 (20.33) | 0.17 (0.06) |
| **KPCA-Tree** | 11.99 (1.23) | 248.0 (44.00) | 0.57 (0.32) | **KPCA-Lasso** | 9.42 (0.70) | 135.00 (20.76) | 0.16 (0.07) |
| **LR** | 3.97 (1.16) | 38.86 (33.60) | 0.77 (0.16) | **RF** | 6.30 (0.81) | 68.52 (17.81) | 0.57 (0.10) |
| **PCA-LR** | 9.19 (0.67) | 131.0 (21.19) | 0.18 (0.08) | **PCA-RF** | 8.81 (0.91) | 134.26 (25.97) | 0.18 (0.10) |
| **KPCA-LR** | 9.42 (0.69) | 135.0 (20.77) | 0.16 (0.07) | **KPCA-RF** | 10.68 (0.96) | 189.56 (31.01) | 0.22 (0.19) |
| **GBDT** | 5.97 (0.84) | 64.82 (17.80) | 0.59 (0.11) | | | | |
| **PCA-GBDT** | 9.15 (0.94) | 144.5 (29.20) | 0.16 (0.11) | **BSDMW** | 3.06 (0.88) | 16.92 (9.95) | 0.89 (0.06) |
| **KPCA-GBDT** | 10.27 (0.85) | 173.8 (27.91) | 0.16 (0.13) | | | | |

This paper visually represents the comparison of the prediction accuracy of different models through boxplots. Figure 2, Figure 3 and Figure 4 show the evaluation of the prediction accuracy of fat content, water content, and protein content under different models, respectively. Among them, (a), (b) and (c) of each picture represent the comparison of prediction accuracy of different models in the non-dimensionality reduction state. (d), (e) and (f) of each picture represent the comparison of the prediction accuracy of different models when performing PCA dimensionality reduction. (g), (h) and (i) of each figure represent the comparison of the prediction accuracy of different models when KPCA dimensionality reduction is performed. The specific results of mean and standard deviation (the values in parentheses in the table are standard deviation values) of the absolute error (AE), mean square error (MSE) and $AR^2$ obtained through simulation experiments between the BSDMW model and other methods are shown in the Table 1, Table 2, and Table 3. Among them, no prefix represents the case of non-dimensionality reduction, PCA represents the case of dimensionality reduction through the PCA method, and KPCA represents the situation of dimensionality reduction through the KPCA method.

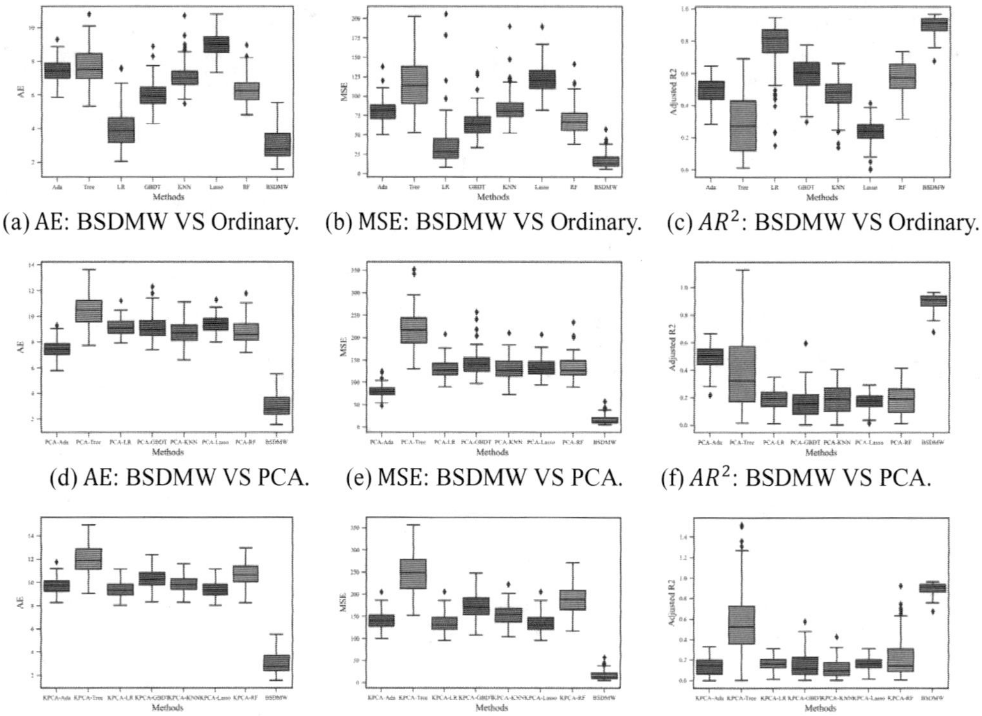

(a) AE: BSDMW VS Ordinary. (b) MSE: BSDMW VS Ordinary. (c) $AR^2$: BSDMW VS Ordinary.

(d) AE: BSDMW VS PCA. (e) MSE: BSDMW VS PCA. (f) $AR^2$: BSDMW VS PCA.

(g) AE: BSDMW VS KPCA. (h) MSE: BSDMW VS KPCA. (i) $AR^2$: BSDMW VS KPCA.

**Figure 1.** Distribution of fitting accuracy of 100 experimental results (fat content of meat samples).

**Table 2.** Fitting accuracy of 100 experimental results (water content of meat samples).

| Method | AE | MSE | AR$^2$ | Method | MSE | AE | AR$^2$ |
|---|---|---|---|---|---|---|---|
| **Ada** | 5.73 (0.41) | 46.03 (6.16) | 0.51 (0.08) | **KNN** | 5.38 (0.51) | 47.65 (8.82) | 0.50 (0.10) |
| **PCA-Ada** | 6.83 (0.54) | 71.90 (11.27) | 0.24 (0.11) | **PCA-KNN** | 6.76 (0.66) | 76.00 (13.85) | 0.22 (0.12) |
| **KPCA-Ada** | 7.46 (0.55) | 80.56 (10.55) | 0.16 (0.09) | **KPCA-KN N** | 7.51 (0.59) | 86.96 (12.76) | 0.13 (0.13) |
| **Tree** | 5.85 (0.83) | 67.52 (17.62) | 0.31 (0.17) | **Lasso** | 2.84 (0.42) | 12.71 (3.55) | 0.86 (0.05) |
| **PCA-Tree** | 8.06 (1.03) | 126.19 (25.86) | 0.37 (0.29) | **PCA-Lasso** | 7.34 (0.45) | 76.51 (10.18) | 0.20 (0.06) |
| **KPCA-Tree** | 9.01 | 140.78 | 0.50 | **KPCA-Lass** | 7.21 | 76.01 | 0.20 |

| | | | | | | | |
|---|---|---|---|---|---|---|---|
| | (0.83) | (21.66) | (0.25) | **o** | (0.46) | (10.10) | (0.06) |
| **LR** | 3.00<br>(0.71) | 22.91<br>(13.76) | 0.76<br>(0.15) | **RF** | 4.94<br>(0.49) | 40.20<br>(7.88) | 0.57<br>(0.09) |
| **PCA-LR** | 7.05<br>(0.47) | 74.50<br>(10.90) | 0.22<br>(0.08) | **PCA-RF** | 6.79<br>(0.68) | 77.47<br>(14.81) | 0.20<br>(0.12) |
| **KPCA-LR** | 7.20<br>(0.46) | 76.02<br>(10.12) | 0.20<br>(0.07) | **KPCA-RF** | 8.02<br>(0.69) | 106.27<br>(16.78) | 0.20<br>(0.12) |
| **GBDT** | 4.67<br>(0.53) | 38.33<br>(8.27) | 0.59<br>(0.09) | | | | |
| **PCA-GBDT** | 7.01<br>(0.66) | 83.29<br>(15.42) | 0.16<br>(0.11) | **BSDMW** | 2.43<br>(0.54) | 10.42<br>(4.28) | 0.89<br>(0.05) |
| **KPCA-GBDT** | 7.70<br>(0.67) | 96.54<br>(15.50) | 0.13<br>(0.10) | | | | |

According to the results of the above charts, we can find that because the BSDMW method proposed in this paper can extract the functional information of the original sequence, and thanks to the data enhancement effect of the stacking framework, the important features of the dataset after the B-spline basis expansion can be effectively enhanced. Therefore, the prediction accuracy of the BSDMW model is higher. In the case of using the original data, the absolute error and the mean square error of the results obtained by the BSDMW method are much smaller than those of the other seven comparison methods, and the average values of $AR^2$ are higher than those of the other seven methods. The standard deviation of these indicators is small. In the case of PCA and KPCA dimensionality reduction data, although when the decision tree model is applied to the protein content of meat samples, the $AR^2$ of the prediction results is slightly higher than that of the BSDMW model, and other indicators of the prediction results of the BSDMW model are better than the other seven comparison models. On the whole, the BSDMW model has an excellent fitting effect when making predictions.

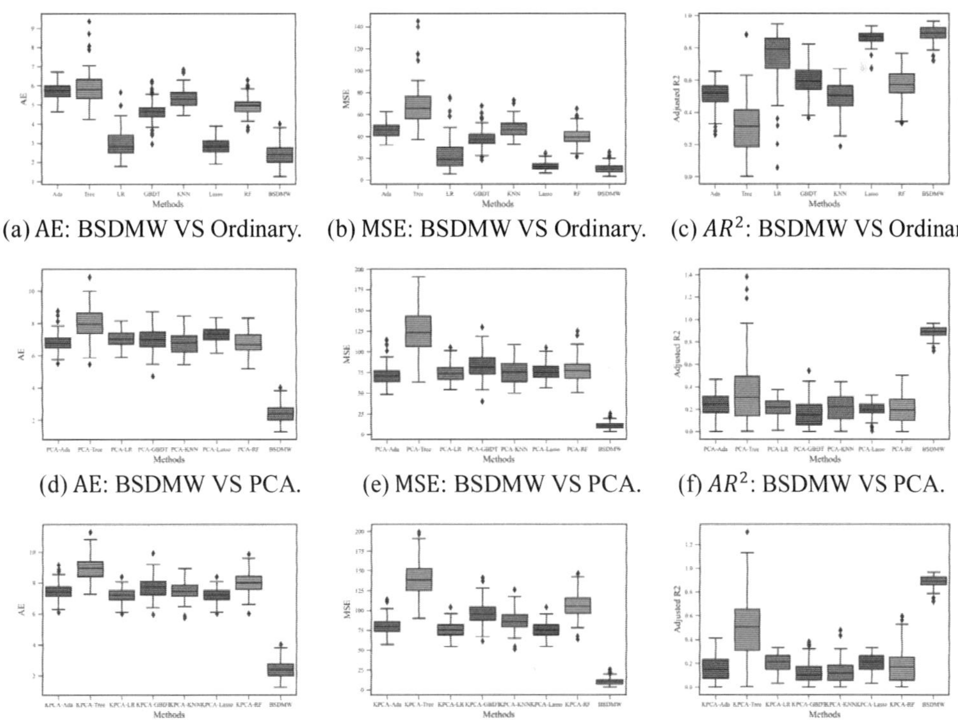


(a) AE: BSDMW VS Ordinary. (b) MSE: BSDMW VS Ordinary. (c) $AR^2$: BSDMW VS Ordinary.

(d) AE: BSDMW VS PCA. (e) MSE: BSDMW VS PCA. (f) $AR^2$: BSDMW VS PCA.

(g) AE: BSDMW VS KPCA. (h) MSE: BSDMW VS KPCA. (i) $AR^2$: BSDMW VS KPCA.

**Figure 2.** Distribution of fitting accuracy of 100 experimental results (water content of meat samples).

Journal of Physics: Conference Series **2813** (2024) 012024 doi:10.1088/1742-6596/2813/1/012024

**Table 3.** Fitting accuracy of 100 experimental results (protein content of meat samples).

| Method | AE | MSE | AR² | Method | MSE | AE | AR² |
|---|---|---|---|---|---|---|---|
| **Ada** | 2.56 (0.17) | 9.17 (1.14) | 0.08 (0.09) | **KNN** | 1.98 (0.20) | 6.24 (1.23) | 0.30 (0.11) |
| **PCA-Ada** | 2.56 (0.17) | 9.17 (1.14) | 0.08 (0.09) | **PCA-KNN** | 2.59 (0.19) | 10.00 (1.38) | 0.15 (0.12) |
| **KPCA-Ada** | 2.61 (0.16) | 9.34 (1.02) | 0.08 (0.07) | **KPCA-KNN** | 2.64 (0.20) | 10.35 (1.456) | 0.17 (0.12) |
| **Tree** | 2.18 (0.27) | 8.70 (1.88) | 0.18 (0.16) | **Lasso** | 2.59 (0.17) | 9.16 (1.15) | 0.02 (0.03) |
| **PCA-Tree** | 3.16 (0.30) | 16.67 (2.56) | 0.87 (0.31) | **PCA-Lasso** | 2.59 (0.17) | 9.18 (1.14) | 0.02 (0.03) |
| **KPCA-Tree** | 3.25 (0.34) | 17.28 (3.01) | 0.94 (0.36) | **KPCA-Lasso** | 2.59 (0.17) | 9.18 (1.13) | 0.02 (0.03) |
| **LR** | 1.25 (0.32) | 3.76 (2.25) | 0.60 (0.20) | **RF** | 1.85 (0.18) | 5.48 (1.07) | 0.39 (0.11) |
| **PCA-LR** | 2.50 (0.16) | 8.70 (1.02) | 0.06 (0.04) | **PCA-RF** | 2.80 (0.25) | 12.54 (2.04) | 0.41 (0.21) |
| **KPCA-LR** | 2.53 (0.16) | 8.88 (1.03) | 0.05 (0.03) | **KPCA-RF** | 2.88 (0.26) | 13.01 (2.04) | 0.46 (0.22) |
| **GBDT** | 1.79 (0.19) | 5.39 (1.12) | 0.40 (0.12) | | | | |
| **PCA-GBDT** | 2.67 (0.23) | 11.39 (1.84) | 0.29 (0.18) | **BSDMW** | 0.92 (0.26) | 1.61 (1.11) | 0.82 (0.11) |
| **KPCA-GBDT** | 2.75 (0.24) | 11.77 (1.86) | 0.32 (0.19) | | | | |

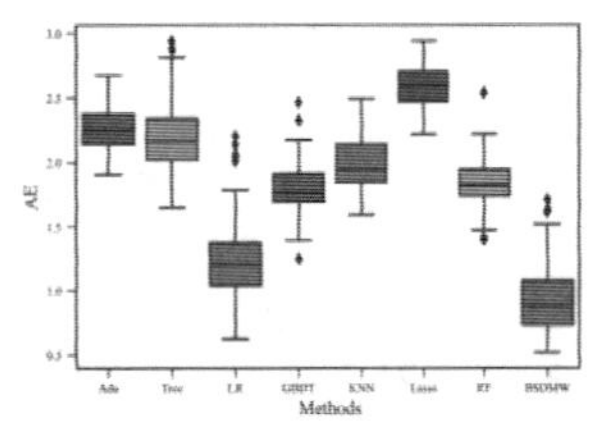
(a) AE: BSDMW VS Ordinary.

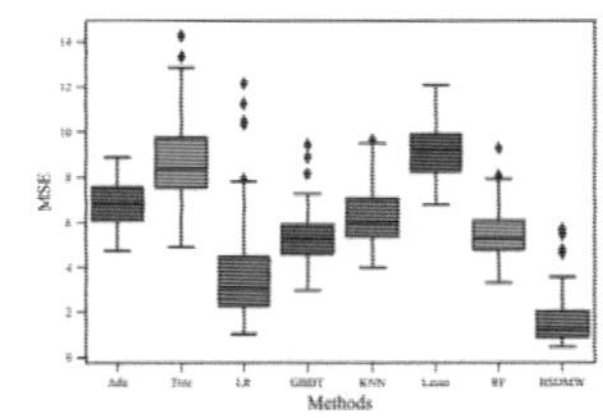
(b) MSE: BSDMW VS Ordinary.

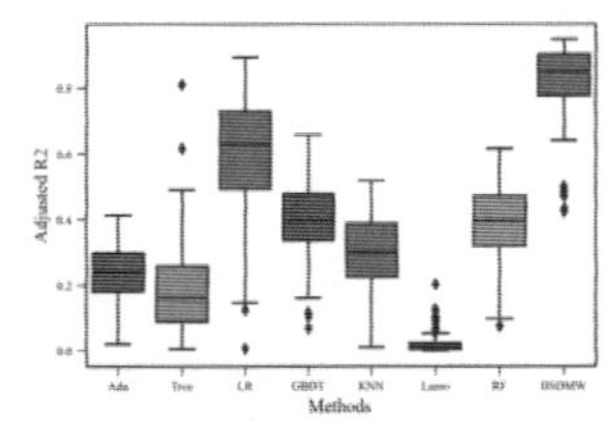
(c) $AR^2$: BSDMW VS Ordinary.

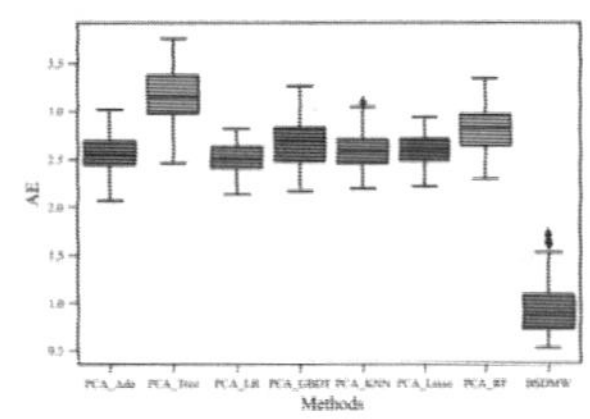
(d) AE: BSDMW VS PCA.

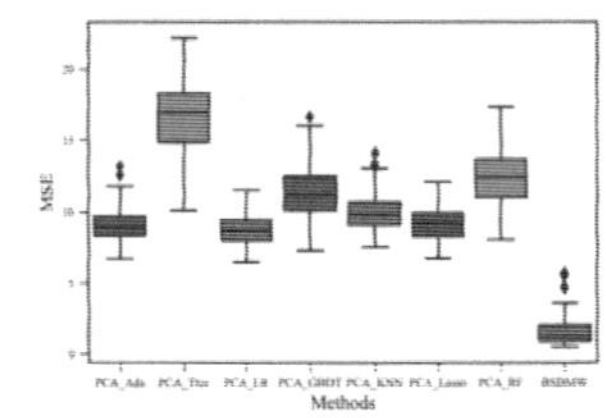
(e) MSE: BSDMW VS PCA.

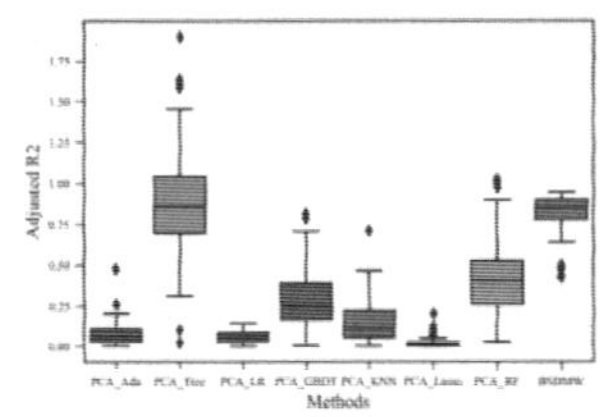
(f) $AR^2$: BSDMW VS PCA.

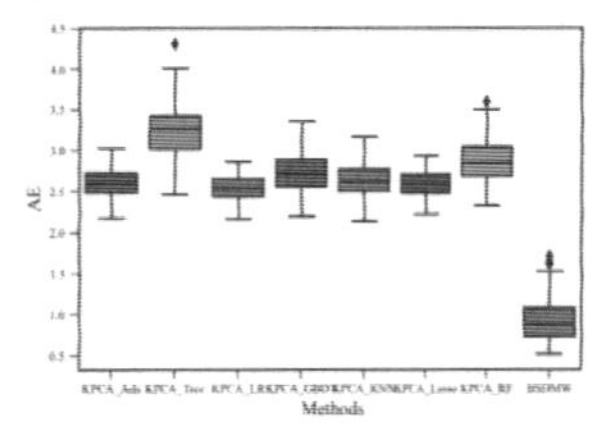
(g) AE: BSDMW VS KPCA.

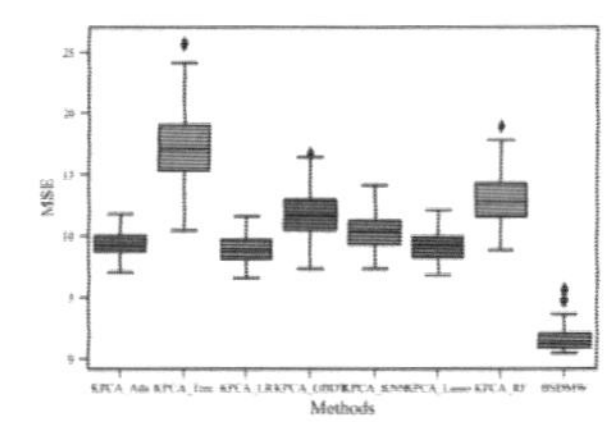
(h) MSE: BSDMW VS KPCA.

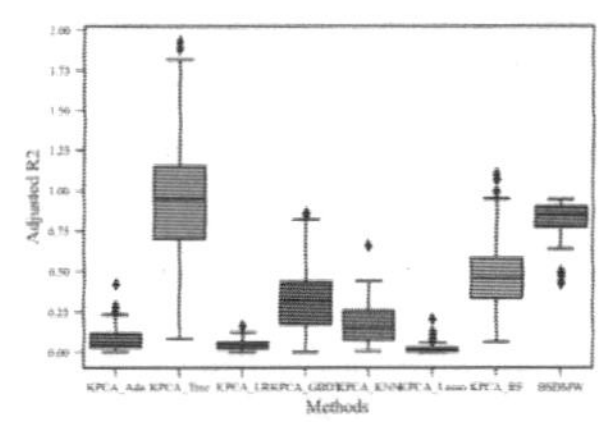
(i) $AR^2$: BSDMW VS KPCA.

**Figure 3.** Distribution of fitting accuracy of 100 experimental results (protein content of meat sample).

## 4. Summary

This paper proposes a time series regression model based on stacking model fusion and weighted bagging framework based on distance correlation coefficient, which has the following advantages. (1) Capturing important features and dimensionality reduction: the model captures the important features in the time series functions to achieve dimensionality reduction for high-dimensional data, thereby optimizing the model's performance; (2) Considering data correlation: the method considers the existence of correlation between time-distance model data, and for some neural network models, avoids the problem of underfitting due to insufficient sample size, which leads to unsatisfactory model effects; (3) Data enhancement through stacking framework: stacking framework is applied to realize data enhancement and further eliminate data noise; (4) Model weighted averaging: the method performs model-weighted averaging based on the degree of importance of the base model, and can adaptively weigh the bias and variance of the prediction model.

After experimental validation, the method used in this paper is found to have higher prediction accuracy and better performance compared with the traditional regression method, with $MSE$, $AE$, $AR^2$ as evaluation indicators. Next, further research will be focused on the following aspects in depth: (1) Model structure: the model in this paper uses the B-sample basis expansion method to fit the function and the distance correlation coefficient as the model importance assessment metric so that the differences in model effects between different basis expansion methods and different weighting metrics can be further explored; (2) Model applications: this paper is applied to time series regression for the content of fat ($y_1$), water ($y_2$), and protein ($y_3$) prediction alone, and further applications will be realized in multiple response regression models (PLS) next.

**References**

[1] Montgomery D C, Peck E A, Vining G G. Introduction to Linear Regression Analysis, 5th Edition [M]. Wiley, 1982.

[2] James G, Witten D, Hastie T, et al. An Introduction to Statistical Learning, with Applications in R [J]. 2021.

[3] Huang M, Wang Q, Jing R, et al. Tropical cyclone full track simulation in the western North Pacific based on random forests [J]. *Journal of Wind Engineering and Industrial Aerodynamics*, 2022:105119.

[4] Nguyen D D, Roussis P C, Pham B T, et al. Bagging and Multilayer Perceptron Hybrid Intelligence Models Predicting the Swelling Potential of Soil [J]. *Transportation Geotechnics*, 2022: 100797.

[5] Xie Z, Liu L, Wang C. Model-Guided Boosting for Image Denoising [J]. *Signal Processing*, 2022: 108721.

[6] Feng J, Liu F, Kuang Y, et al. Prediction of nosocomial pneumonia after brain injury: a PCA-logistic regression model based on machine learning [J]. *Chinese Journal of Clinical Neurosurgery*, 2019, v. 24 (01):40-43.

[7] Xu W, Tang J, Xia H, et al. Prediction method of dioxin emission concentration based on PCA and deep forest regression [C]. *Proceedings of the 40th China Control Conference*, 2021:6.

[8] Zhou M, Zeng X, Chen A. Deep Forest Hashing for Image Retrieval [J]. *Pattern Recognition*, 2019, 95.

[9] Pérez-Montalvo E, Zapata-Velásquez M E, Benítez-Vázquez L M, et al. Model of monthly electricity consumption of healthcare buildings based on climatological variables using PCA and linear regression [J]. *Energy Reports*, 2022, 8: 250-258.

[10] Kdcp A, Gkw B, Rh C, et al. Study of Concentration Dependent Curcumin Interaction with Serum Biomolecules using ATR-FTIR Spectroscopy Combined with Principal Component Analysis (PCA) and Partial Least Square Regression (PLS-R) [J]. *Vibrational Spectroscopy*, 2021.

[11] Rb A, Dc A, Kja B. A possibilistic regression based on gradual interval B-splines: Application for hyperspectral imaging lake sediments [J]. *Information Sciences*, 2020, 510:183-204.

[12] Bignardi A B, Faro L E, Santana M L, et al. Bayesian analysis of random regression models using B-splines to model test-day milk yield of Holstein cattle in Brazil [J]. *Livestock Science*, 2012, 150(1-3):401-406.
[13] Hayashi K, Hayashi M, Reich B, et al. Functional data analysis of mandibular movement using third-degree b-spline basis functions and self-modeling regression [J]. *Orthodontic Waves*, 2012, 71(1):17-25.
[14] Soh C G, Zhu Y. A sparse fused group lasso regression model for fourier-transform infrared spectroscopic data with application to purity prediction in olive oil blends [J]. *Chemometrics and Intelligent Laboratory Systems*, 2022, 224:104.

# AUTHOR INDEX